International Military Operations in the 21st Century

This book examines the challenges that military forces will face in multinational operations in the twenty-first century.

Expanding on Rupert Smith's *The Utility of Force*, the volume assesses the changing parameters within which force as a political instrument is ultimately exercised. By analysing nine carefully selected mission types, the volume presents a comprehensive analysis of key trends and trajectories. Building upon this analysis, the contributors break the trends and parameters down into real and potential tasks and mission types in order to identify concrete implications for military forces in future multinational operations.

The context of military intervention in conflicts and crises around the world is rapidly evolving. Western powers' shrinking ability and desire to intervene makes it pertinent to analyse how the cost of operations can be reduced and how they can be executed more intelligently in the future. New challenges to international military operations are emerging and this book addresses these challenges by focusing on three key areas of change: an increasingly urbanised world; the changing nature of missions; and the commercial availability of new technologies. In answering these questions and embracing some of the insights of the growing field of future studies, the volume presents an innovative perspective on future international military operations.

This book will be of much interest to students of international intervention, military and strategic studies, war and conflict studies, security studies and international relations in general.

Per M. Norheim-Martinsen is senior research fellow at the Institute for Defence Studies in Oslo, Norway and author of *The European Union and Military Force* (2013).

Tore Nyhamar is senior researcher at the Norwegian Defence Research Establishment (FFI) and has a PhD in Political Science from the University of Oslo.

Cass Military Studies

International Military Operations in the 21st Century

Global trends and the future of intervention

Edited by
Per M. Norheim-Martinsen
and Tore Nyhamar

Routledge
Taylor & Francis Group

LONDON AND NEW YORK

First published 2015
by Routledge

2 Park Square, Milton Park, Abingdon, Oxon OX14 4RN
711 Third Avenue, New York, NY 10017, USA

Routledge is an imprint of the Taylor & Francis Group, an informa business

First issued in paperback 2016

Routledge is an imprint of the Taylor & Francis Group, an informa business

British Library Cataloguing-in-Publication Data
A catalogue record for this book is available from the British Library

Library of Congress Cataloging-in-Publication Data
International military operations in the 21st century : global trends
and the future of intervention / edited by Per M. Norheim-
Martinsen and Tore Nyhamar.
    pages cm. – (Cass military studies)
    Includes bibliographical references and index.
    1. Combined operations (Military science)–Forecasting.
    2. Multinational armed forces–Forecasting. 3. Intervention
    (International law) 4. War–Forecasting. 5. Twenty-first century–
    Forecasts. I. Norheim-Martinsen, Per M., editor of compilation.
    II. Nyhamar, Tore, editor of compilation.
    U260.I54 2015
    355.4'6–dc23   2014033645

ISBN 978-1-138-81915-3 (hbk)
ISBN 978-1-138-69441-5 (pbk)

Typeset in Baskerville
by Wearset Ltd, Boldon, Tyne and Wear

Contents

Contributors

Alexander William Beadle has been a research fellow at the Norwegian Defence Research Establishment (FFI) since 2011. He received his BA in War Studies and MA in Conflict, Security and Development from King's College London. His research has primarily focused on violence against civilians in contemporary conflicts, and he has developed planning guidance for the protection of civilians in military operations. He is currently leading a project on future military operations for the Norwegian Armed Forces.

Sverre Diesen is a retired general and has been a chief researcher at FFI since 2012. He served in a number of staff and command appointments in the Norwegian armed forces, becoming Chief of Defence 2005–2009. He has published two books on strategy and the future development of the Norwegian military, as well as a number of articles in military journals. He holds an MSc in civil engineering from The Norwegian University of Technology and Science.

Robert Egnell (PhD King's College, London) is a Visiting Professor and Director of Teaching with the Security Studies Program at Georgetown University. He is currently on leave from a position as Associate Professor at the Swedish National Defence College. He is co-author (with David Ucko) of *Counterinsurgency in Crisis: Britain and the Challenges of Modern Warfare* and author of *Complex Peace Operations and Civil–Military Relations: Winning the Peace* (Abingdon, UK: Routledge, 2009), and *Gender, Military Effectiveness and Organizational Change: The Swedish Model* (Basingstoke, UK: Palgrave 2014).

Iver Johansen has been a chief researcher at FFI since 1988. He holds a degree in political science from the University of Oslo (1987). Between 2009 and 2012, he conducted research on land forces development and long-term force planning for the Norwegian Army. He currently leads research activities on the nature of special operations and on the future development of Norway's special operations forces.

Siw Tynes Johnsen is a research fellow at FFI. She received her BA in Comparative Politics from the University of Bergen, and her MA in International Relations from the University of Kent's Brussels School of International Studies. She worked as a media analyst in NATO Headquarters during Operation Unified Protector (Libya). Her current research at FFI focuses on the role of the cyber domain in international military operations.

Stian Kjeksrud is a senior researcher at FFI, a PhD candidate at the University of Oslo and currently affiliated with the University of Cape Town. He is a former officer with experience from operations in Afghanistan, Kosovo, Macedonia and Lebanon. His research is on contemporary UN peacekeeping operations in Africa and the role of military force to protect civilians, publishing 'United Nations Stabilization Operations: The DRC and Mali' with Lotte Vermeij, in *UN Peacekeeping Doctrine in the Post-Brahimi Era* (Oxford: Oxford University Press).

Guro Lien has been working as a researcher at the Norwegian Defence Research Institute since 2008. She has an MSc in Comparative Politics from the London School of Economics and Political Science. Her research interests are within international operations, and in particular security sector reform and military assistance. She has published a number of reports, articles and a book chapter on these topics.

Per M. Norheim-Martinsen is a senior research fellow at the Institute for Defence Studies in Oslo. He received his PhD in International Studies from the University of Cambridge. He is a former officer, with experience from operations in Lebanon. He has published extensively on European security and international military operations, including *The European Union and Military Force* (Cambridge: Cambridge University Press, 2013). His current research focuses on the transformation of European military forces.

Tore Nyhamar (PhD University of Oslo) has been a senior researcher at FFI since 2001. He obtained his Doctorate from the Department for Political Science at the University of Oslo, where he held various positions from 1989 to 2000. He has been Director of research on FFI's work on international operations since 2008. His research focuses on counterinsurgency, small states in international operations and military doctrines.

David H. Ucko (PhD King's College, London) is an associate professor at the College of International Security Affairs, National Defense University, and an adjunct fellow at the Department of War Studies, King's College London. He is co-author (with Robert Egnell) of *Counterinsurgency in Crisis: Britain and the Challenges of Modern Warfare* (New York: Columbia University Press, 2013) and author of *The New Counterinsurgency Era: Transforming the U.S. Military for Modern Wars* (Washington DC: Georgetown University Press, 2009).

Preface and acknowledgements

This book has been a long time in the making. But the result, we hope, is a truly different perspective on international military operations in the twenty-first century, one that will prove helpful to practitioners and policymakers, and thought-provoking for anyone interested in issues of international intervention today. The project began in 2011 when the Challenges in Peace and Stabilisation Operations group at the FFI was challenged to say something about what future international operations might look like, as part of the institute's support activities for the Norwegian Armed Forces' long-term defence planning. The group then consisted (in alphabetical order) of Stian Kjeksrud, Anders Kjølberg, Per M. Norheim-Martinsen, Tore Nyhamar and Jacob Aasland Ravndal. Actually, the question we most often get is 'where will our armed forces be engaged next'. We do not particularly care for this question, as our research does not tell us what future events have in store for Western decision-makers. We had recognised even prior to our first brainstorming session that our response would have to be something other than our best guess on where Norway would engage next. Indeed, the resulting FFI-Report 2011/01697 *Fremtidens internasjonale operasjoner [Future international operations]* by Norheim-Martinsen *et al.* explains why the question 'where will be next' is misguided. The report was co-published with FFI-Report 2011/01667 *Trender, scenarioer og sorte svaner – utfordringer for fremtidens landmakt [Trends, Scenarios and Black Swans – Challenges for Future Land Power]* by Norheim-Martinsen, in which the methodology for the present volume was first laid out.

The former report was presented at an FFI seminar in Oslo with Lt Gen Robert Mood, who headed the United Nations Supervision Mission in Syria (UNSMIS) in 2012, and Prime Minister Erna Solberg, then leader of the Conservative Party, commenting. During the stimulating discussion, Solberg observed: 'I hear what the military is saying about what kind of international operation we ought to engage in, but it is not easy to be a politician, sometimes things just happen, and it is necessary to act.' This is the premise of the book in a nutshell. International military operations are rarely the result of a neat planning process, where the capabilities we want to send are the capabilities needed, the timeframe is ideal, and the challenges what we thought they would be when the decision to deploy was made. We cannot stress enough that this book is not an attempt to *predict* what the future of international military operations will look like. But we hold that the option of international use of force to confront crimes against humanity and manage crises that threaten local, regional and global security is one that will be, and should be, open in the future also. This requires constant adaptation of national and international military structures and capabilities to a changing conflict environment. With this book we want to encourage foresight in planning by raising the awareness, among practitioners, policymakers and students alike, of what new challenges this changing conflict environment may pose in the future.

The present volume builds and improves on the first report in two ways. First, the number of mission types is expanded. Cyber operations, Special Forces operations, United Nations (UN) operations and protection operations have been added. Second, it was possible to cover each mission in depth, and further develop and utilise the conceptual framework. We were so fortunate to be able to expand the team and include the expertise of Alexander William Beadle, Gen (ret.) Sverre Diesen, Robert Egnell, Iver Johansen, Siw Tynes Johnsen, Guro Lien and David Ucko. In addition to being a pleasure to work with, they all contributed in so many ways to making this book a truly joint effort. We also take the opportunity to express our gratitude to Lene Ekhaugen, Lt Col Egil Daltveit, Anne Kjersti Frøholm, Kirsten Gislesen, Anne Lise Hammer, Maj Tor-Erik Hanssen, Torbjørn Kveberg, Lt Col Kjell Pedersen, Ragnhild Siedler, Henning Søgaard, Anders S. Vaage and Lt Col Palle Ydstebø.

Thanks also to two anonymous referees who understood what we tried to do and offered constructive advice to keep us on track. We appreciate the unwavering support from FFI Head of the Analysis Division Espen Skjelland throughout the process, and from Project Leader Frode Rutledal for his patience at the end. Last but not least a heartfelt thanks to Tom Major at the FFI. Without his timely intervention way beyond the call of duty, the book may well have remained uncompleted.

Tore Nyhamar and Per M. Norheim-Martinsen
Oslo

Abbreviations

AMISOM	African Union Mission in Somalia
AU	African Union
BOPE	Special Police Operations Battalion (Brazil)
CIA	Central Intelligence Agency (United States)
COIN	Counter-insurgency
DDOS	Distributed Denial of Service
DPKO	UN Department of Peacekeeping Operations
DRC	Democratic Republic of Congo
ECOWAS	Economic Community of Western African States
EU	European Union
EW	Electronic Warfare
FARC	*Fuerzas Armadas Revolucionarias de Columbia*
FARDC	Congolese Army
FIB	Force Intervention Brigade
FFI	Norwegian Defence Research Establishment
GSG 9	German Police Counterterrorist Unit
IC	International Community
ICC	International Criminal Court
ICRC	International Committee of the Red Cross
IDF	Israeli Defence Force
ISAF	International Security Assistance Force
LRA	Lord's Resistance Army (Uganda and around)
MINUSMA	UN Multidimensional Integrated Stabilization Mission in Mali
MINUSTAH	UN Stabilisation Mission in Haiti
MONUSCO	The United Nations Organization Stabilization Mission in the Democratic Republic of the Congo
NTC	National Transitional Council (Libya)
OUP	Operation Unified Protector (NATO)
OMLT	Operational Mentoring and Liaison Team
P–5	Permanent Five Members of the Security Council
PMC	Private Military Company
R2P	Responsibility to Protect
SACEUR	Supreme Allied Commander Europe

SAS	Special Air Service (British)
SOCOM	United States Special Operations Command
SPLA	Sudan People's Liberation Army
UN	United Nations
UNHCR	UN High Commissioner for Refugees
UNISFA	UN Interim Security force for Abyei
UNMISS	UN Mission in the Republic of South Sudan
UNMIX	Imaginary Peacekeeping Force
UNSC	UN Security Council
UNSMIS	UN Supervision Mission in Syria
UPP	Pacification Police Units, Brazil
US	United States
US CERT	US Computer Emergency Response Team
USCYBERCOM	US Cyber Command

1 Introduction

Trends and scenarios in international military operations

Per M. Norheim-Martinsen

The context in which international military operations are carried out is constantly changing, as is the nature of war itself. The purpose of this book is nothing less than an attempt to understand how current trends, and the trajectories they may take, will affect tomorrow's military operations. We are fully aware of the pitfalls of trying to look into the future – many are those who have failed to do so in the past. But we hold that by thinking about the future we will be better prepared to deal with both the expected and the unexpected.

The book is ultimately about finding the utility of force in a modern world. The reference to General Sir Rupert Smith's opus magnum from 2005 is, of course, no coincidence (Smith 2005). Regarded by many as the 'most important book on war for over a century and a half' (Gow 2006, p. 1152), Smith offers an insightful account of the changing nature of military strategy, operations and tactics. But where Smith starts his analysis by asking what our desired end for using force is, we ask how the parameters in which force as a political instrument is ultimately exercised will affect its use in the future. In light of demographic and economic shifts, how can operations be carried out cheaper or smarter? How will the fact that half of the world's population already lives in cities affect future operations? How will emerging powers like China, India, Brazil and Turkey affect how future peacekeeping operations are carried out? What missions must military forces expect to carry out in the future? And how will new technologies affect future operations?

By addressing these questions, the individual chapters of the book aim to break down general trends and trajectories into tasks and mission types in order to identify concrete implications for military forces in tomorrow's international military operations. As such, the book has a more specific aim than some of the excellent books on the market, including Smith's, which address the changing character of war from a more general perspective (Barnett 2004, Gray 2005, Haug and Maao 2011, Strachan and Scheipers 2011). It also takes a more explicitly forward-looking perspective than much of the recent literature on peace and stabilisation operations, which is for the most part based on historical analysis of past and present operations (Daniel *et al.* 2008, Paris and Sisk 2009, Mayall and de Oliveira 2011). By embracing instead some of the insights of the growing field of future studies, and merging them with traditional scenario thinking, the individual authors of this book are able to offer fresh insights and perspectives on some of the challenges that military forces may have to face in the future.

This introductory chapter first discusses briefly some of the limits and potentials of studying the future, and outlines the methodological basis for the book. It explains the difference between forecast and foresight, and points towards how future studies may complement long-term defence planning. The chapter then offers an overview of key global trends and trajectories, and outlines how they relate to future international military operations. The analysis covers trends in areas like demographics, economics, technology and geopolitics and leads into nine scenarios or generic mission types, which are then analysed in each of the remaining chapters of the book. A key objective of the volume has also been to come up with a comprehensive set of generic parameters for analysing emerging tasks and requirements arising from each mission type. These are set out and discussed in Chapter 2.

Future studies and long-term defence planning

Coming from a research community intimately involved in defence planning on both the national and international (NATO) levels, we see the need for a different approach to long-term defence planning than the traditional models applied today. Modern defence planning is usually based on scenarios. They are, as such, studies of what are perceived to be likely but unwanted future events, and ways of managing these events. However, there is no guarantee that other events, which are not planned for, will occur. In fact, unexpected events represent rather the rule than the exception, a fact of life that has earned future studies a rather dodgy reputation, especially in academic circles. However, we hold that this understanding is dated, and ultimately misses what future studies are really about.

Future studies are often criticised for trying to predict the future – and, in the end, for usually getting it wrong. But the purpose of future studies is *not* to predict. Rather, future studies are about improving our capacity to manage the *uncertainty* that inevitably defines the future – it is about thinking 'outside the box'. To be able to say at some point down the line that our analysis of the future corresponded with the actual events is not really the key. The aim of future studies is to make sure that decision makers and planners do not get locked in to rigid and/or faulty interpretations of continuity and change, and therefore make rash decisions, or hesitate to make decisions, when unexpected or expected events occur. But, before moving on to discuss some of the limits and potentials of future studies, it is necessary to clarify what we mean by 'the future'.

There is no set timeframe for the book, insofar as the time horizon for the studies on which our analysis of global trends draw varies. For example, in demographics, it is not unusual to have a 30–50-year-perspective. In technology studies, it is typically less than 25 years. In demographics, one is fairly certain how many people are born in a given year, providing a relatively firm basis for predicting how the population will develop in the future. In technology studies, the time horizon is shorter because breakthroughs and rapid change lead to significantly greater uncertainty. In any case, giving an exact date for how far into the future one would wish to see may imply a degree of certainty that simply does not exist and may thus be misleading.

Instead, it is more fruitful to contrast our perspective with the timeframe and methodology of traditional long-term defence planning. The book is intended to complement this process, which in most Western states is carried out in four-year cycles leading up to major reviews. In our experience, it is difficult to include upcoming needs for change in this process, and this is a problem which is shared with colleagues in most NATO countries. Some countries, such as Norway, have moved towards a continuous demand-driven planning process, which takes place simultaneously with implementation, to counter this tendency. However, a problem with this approach is that it tends to be too reactive and does not encourage a more fundamental analysis of continuity and change at regular intervals (see Norheim-Martinsen *et al.* 2011). A key objective of future studies is precisely to 'look beyond the horizon', and to illustrate to decision makers how global trends and alternative scenarios may pose both expected and unexpected challenges in a longer perspective. Accordingly, the timeframe of the book stretches beyond – but does include – the 4–12-year-perspective adopted in long-term defence planning, but it takes a different perspective and methodology in order to complement this approach.

Looking into the crystal ball – limits and potentials of future studies

As stated above, the purpose of future studies is not to try to *predict* the future, but rather to improve our capacity to manage *uncertainty*. This is also why contemporary future studies literature tends to prefer the term 'fore*sight*' to 'fore*cast*'. However, despite the usual caveats included in the forewords and introductions of the numerous trend reports, books and future blurbs that are published each year, they cannot escape the criticism that they more often than not fail to predict the intensity and direction even of trends that come across as fairly certain, or fail to predict the important events that in hindsight come across as perfectly predictable (see e.g. Gardner 2010). The latter are often referred to as ruptures, strategic surprises or shocks, or in more popular terms, 'black swans'.

In his 2007 international bestseller *The Black Swan: The Impact of the Highly Improbable*, Nicholas Nassim Taleb claims that almost everything in the world, from major scientific discoveries to historical events, or the directions our own lives take, are due to 'black swans' – rare, unexpected and completely unpredictable events, which, nevertheless, come across as completely rational when seen in hindsight. He cites the development of the Internet, the computer, World War I and the 11 September 2001 terrorist attacks as examples of such events. Taleb's key message is that even if we are able to predict certain trends with some degree of probability, we cannot predict the really important, but rare, events that will have the greater impact on our lives. The fact that they are, indeed, rare is what makes them fall outside the remits of human imagination – human psychology is biased towards what is known to us. This does not imply that the events themselves are rare, merely that we are unable to predict when, where and how the next black swan will strike.[1] Hence, having better or more statistics does not necessarily improve our ability to predict the future. Likewise, trends and projections are nothing more than future expectations based on developments over time – or like driving by looking in the rear-view mirror.

The black swan problem is an important corrective to a growing belief that risk analysis and management may put us in a position to identify and remove, or at the very least reduce, any threat. Over the latter years, there has emerged a virtual industry dedicated to future studies, and risk analysis in particular. For example, the United Nations University Millennium Project Handbook, which is updated regularly, describes more than 25 different future analysis methods (see Glenn and Gordon 2009). None of them is able to offer certainty about the future. Yet all risk analysis is based on calculations in which risk is a function of the probability that a threat will occur, and the consequences of it occurring. A precondition for all risk analysis, therefore, is that any event can be predicted with some degree of precision. History tells us that is not the case. Rather, the increased belief that risk can be managed, and the corresponding development of new methods aimed at increasing the accuracy of future studies, may quickly become a form of collective self-deception. The question is then: why should this book concern itself with the future of international military operations?

First, although we cannot predict the future, it makes perfect sense to anticipate that *something* is going to happen, and to try to limit the negative consequences of potential future threats. The paradox of military power is that it is most effective when it is *not* used. Rather, it is the existence or threat of force, and the corresponding knowledge about the inevitable costs of going to war, that will ensure its effectiveness. We hold that the option of international use of force to confront crimes against humanity and manage crises that threaten local, regional and global security is one that will be, and should be, open in the future also. This requires constant adaptation of national and international military structures and capabilities to a changing conflict environment. With this book we want to encourage foresight in planning by raising the awareness, among practitioners, policymakers and students alike, of what new challenges this changing conflict environment may pose in the future.

Second, thinking actively about the future may feed into scenarios or mission types other than those planned for today. Scenarios represent a key tool for states and international security organisations to test and prepare military structures and capabilities for the kind of tasks they are likely to (have to) carry out. They make it possible to prioritise and make decisions under conditions of uncertainty. However, it is important that military planners, when they develop their scenarios, do not rule out the kind of tasks and missions that at present come across as less likely. As such, future studies and scenarios may mutually reinforce each other by contributing towards robust structures and capabilities, while, at the same time, ensuring that future studies are made relevant for long-term defence planning. The identification and analysis of nine different mission types against five generic parameters, which we argue in Chapter 2 will define emerging tasks and requirements for future forces, is meant as a modest first step in this regard.

Third, despite the usual criticism launched against future studies, that they tend to get the future wrong, they also often get it right. Conservative analyses of trends and trajectories within isolated areas, such as economics or demographics, may, in fact, prepare the ground for good decisions, and make sure that the proper precautions are taken. Black swans are admittedly all about the major, course-changing events, but planning is also about the small-to-medium events. It is, for example, unwise not to start revising expensive pension and welfare schemes when we expect that the proportion of older people in society will rise. Trends analyses may counter tendencies towards near-sightedness in defence planning by showing how even conservative prognoses may incur completely new security challenges over time. By actively exploring how major global trends, which are not at the outset connected to security and defence, may affect the use of international military force in the future, we want to contribute towards countering such tendencies.

Finally, future studies and trends analyses are a constant reminder that the world is changing. Such activities stimulate alternative thinking and contribute towards avoiding so-called 'cognitive closure' – i.e. human desire to eliminate ambiguity and arrive at definite conclusions. In our search for certainty, we tend to stick with what we know, or what has already happened (see e.g. Webster and Kruglanski 1997).[2] It is, for example, an historical fact that armies prepare for the last war. Or in the famous words of Sir Basil Liddell-Hart: 'The only thing harder than getting a new idea into the military mind is to get an old one out'. Studying the future contributes in itself towards nurturing a collective awareness that unexpected events will occur. By not thinking about the future we succumb to the false idea that the world remains the same.

To sum up, the purpose of the book is not to predict but to provide fore*sight*, as opposed to fore*cast*. Accordingly, we make no attempt to predict what trends or mission types are most likely in the future, nor assign probabilities to the mission types. In this sense our approach differs from, for example, the Correlates of War project (see www.correlatesofwar. org/) or any other approaches aiming to improve statistical predictions regarding the likelihood of war (see e.g. Hegre *et al.* 2013). Instead, foresight handles uncertainty by spelling out a number of scenarios rather than narrowing down to only those that are considered likely at present. It is sufficient that a scenario is plausible. Also, the sample of scenarios or mission types does not need to be exhaustive to provide foresight. It only needs to find enough interesting new challenges to be helpful.

With this aim in mind, we arrived at the nine mission types analysed in this book by considering two ways in which new challenges for the international use of military force might arise. The first is how present mission types will evolve in a conflict environment shaped by global trends. The second is how the global trends may give rise to completely new mission types. We shall return to what the nine mission types are shortly, but only after we outline what we expect to be key global trends and explain how they may shape international military operations in the future.

Demographics of discord – the shrinking of the West

Changing demographics represents a first set of trends that directly or indirectly will affect the conduct of international military operations in the future. Some commentators have even begun to talk about a new 'population bomb' (Goldstone 2010). The numbers speak for themselves.

Given expected growth rates, the world's total population is expected to stabilise at somewhere between 7.7 and 11.2 billion people by 2050 (United Nations Population Reference Bureau, www.prb.org/). But 90 per cent of the population growth will take place outside the Western world. Since 2007, more people have been dying than have been born in Europe. Russia is losing some 0.5 per cent of its population every year. It has been remarked that, pandemics apart, the Russian situation lacks historical precedents (Howe and Jackson 2008, p. 7). In contrast, China's population is expected to grow to 1.4 billion but will be surpassed by India, whose population will grow to 1.8 billion by 2050. The United States' population will grow to 400 million people, while the total population of the Western world will be around one billion people. In 2050, this will make up only one-tenth of the world's total population, compared to just below one-seventh today.

The consequences of this demographic shift are serious enough. But the West's relative decline is also reinforced by a second megatrend: a dramatic ageing of the population in this part of the world. At present, the proportion of people over 60 years old in the United States (US), Canada and the European Union (EU) is 15–22 per cent. But, as the post-war baby boomers are nearing retirement and life expectancy increases, the proportion of people over 60 will increase to 30 per cent. At the same time, the number of people of working age will go down, due to declining birth rates. In some countries, the consequences of these demographic shifts will be harder felt than in others: South Korea will, despite an expected total population loss of only 9 per cent (from 48.3 to 44.1 million) by 2050, experience a 36 per cent decline of its working age population, while its proportion of people over 60 will grow by 150 per cent. As a result, South Korea's working age population will barely outnumber those over 60 years old. Europe will lose some 24 per cent of its working age population, while the proportion of those over 60 will increase by 47 per cent. It is also worth remarking that China will be facing similar prospects (from 12 per cent over 60 today to 30 per cent in 2050), which will, among other factors, have severe consequences for its economic growth. India is again the demographic winner but faces a redistribution problem, as the majority of its population growth will take place in the rural, poor north (US Government 2010, p. 15).

The ageing of the West is, in turn, reinforced by a third demographic megatrend: roughly nine out of ten children under the age of 15 are growing up in the developing world. On the one hand, this emerging 'youth bulge' may provide the basis for improved economic growth in countries like Turkey, Iran, the Maghreb states (Morocco, Algeria and Tunis), Colombia, Costa Rica, Chile, Vietnam, Indonesia and Malaysia (US Government 2008, p. 22). On the other, it may lead to a wave of 'angry youth' if state authorities fail to provide them with opportunities, including education, jobs, income and the prospects of a worthy life. Of the world's total population growth towards 2050, 70 per cent will take place in countries which are classified by the World Bank as low income countries (i.e. average income below US$3.855 in 2008), such as Afghanistan, Pakistan, the Democratic Republic of Congo (DRC), Ethiopia, Nigeria and Yemen (US Government 2008, p. 22; Goldstone 2010).

Finally, the world is urbanising at an unprecedented rate (UN-HABITAT 2008, Goldstone 2010). Today, more than 50 per cent of the world's population is living in cities, a share that is expected to grow to 70 per cent by 2050. This trend is particularly strong in Asia and Africa, which contain most of the world's 19 megacities, i.e. cities with more than ten million people. In sub-Saharan Africa, more than one billion people are expected to live in cities (from a proportion of 35 per cent today to 67 per cent in 2050). The total number of megacities in the world is expected to increase to 25 as early as 2025, but only one of them (Paris) will be in the Western world (US Government 2008, p. 23). From an economic point of view, the concentration of people in urban centres is expected to have positive effects, insofar as roughly 65 per cent of all economic activity and 85 per cent of all innovation take place in cities (UN-HABITAT 2008). But there is a risk that urbanisation will have a destabilising effect upon many countries. The urbanising states in sub-Saharan Africa in particular are characterised by fast growing cities, urban sprawl, the growth of slums and massive inequalities between rich and poor. These environments are prone to rising violence and crime. They are also identified by lack of management and control, which, in turn, may open them up to radical groups, terrorist networks and organised criminals gaining a foothold and being able to operate unchecked (Kilcullen 2013).

All of the abovementioned demographic trends will give rise to new challenges in and for international military operations. Ageing populations in the Western world will have consequences for recruitment to national military forces. Western states already have difficulties sustaining large military deployments overseas, due to shrinking armies and defence budgets (see below). Moreover, shrinking youth rates may increase the threshold for Western states to engage in military operations, given the economic and human costs that the use of armed force inevitably carries. Other states, especially in the Middle East and sub-Saharan Africa, which will often have youth rates above 50 per cent, may, in turn, have a lower threshold for engaging in armed conflict (US Government 2010, p. 15). Among the top ten troop-contributing countries to UN peacekeeping operations, there are no Western states. States like Bangladesh, Pakistan, India, Nigeria, Egypt, Nepal and Jordan continue to bear the greatest burden, while a state like China has become the largest troop contributor among the permanent members of the UN Security Council and the fourteenth largest contributor altogether. Emerging powers, such as Brazil, are moving in to take the lead in new UN operations (see discussion below).

Many of the challenges already seen in today's operations will be exacerbated. Ethical dilemmas in dealing with child soldiers will continue to pose a challenge to intervening forces, especially in countries in Africa with extremely youthful populations. The dividing line between soldiers and civilians will become ever more blurred, as operations move from rural to urban conflict environments. The urbanisation trend of the last 20 years has made it increasingly harder to avoid military operations in urban areas, such as in Beirut (1982), Los Angeles (1992), Mogadishu (1993), Gaza (2009), Grozny (1995, 2000), Baghdad (2003) and Fallujah (2004) (Vautravers 2010, p. 439). In the future, the danger of vulnerable cities collapsing completely, or being subject to massive systemic breakdown, as a result of natural disasters, energy shortages, lack of law and order etc. will have wide security and humanitarian consequences. The need to be able to operate in urban environments will have vast implications for Western militaries when it comes to tactics, doctrine, equipment and training. Tomorrow's megacities may typically contain small pockets of heavily armed opposing forces, lawless areas under the control of criminal gangs, but also areas with relative stability, often in the immediate vicinity of each other. International forces may, therefore, have to carry out tasks such as community policing and more demanding combat operations simultaneously and within the same theatre of operations. The UN operation MINUSTAH (*Mission des Notions Unies pour la Stabilisation en Haiti*) in Haiti showed how military forces might be a blunt yet effective instrument for confronting excessive urban violence. Similar-type operations have been carried out in the favelas of Rio de Janeiro, which illustrates how emerging powers like Brazil, which has been lead nation for MINUSTAH since 2004, are developing the type of experience and capabilities needed in tomorrow's urban military operations. This is covered in depth by Per M. Norheim-Martinsen in Chapter 7 of the book.

Finally, as a result of urbanisation, most refugees come from cities, and therefore seek out other cities for protection. Protection of civilians is a main objective in all UN peace operations today (see also below). The challenges of protection in contemporary conflicts became all too apparent in the 2010 international military campaign in Libya, in which the limitations of air power were revealed as the warring parties retreated to the cities. Protection of civilians in international military operations is covered in depth by Alexander W. Beadle in Chapter 9.

Economic trends – spending smarter or going out of business

Economic developments are closely connected to the demographic trends above. Here too the main trend is a relative weakening of the West due to lack of workers, fewer consumers and low expected economic growth in the years to come. To put recent developments in perspective: in the period after the industrial revolution, the US, Canada and Europe increased their share of global GDP from 32 per cent in 1900 to 68 per cent in 1950. Between 1950 and 2003, this share fell to 47 per cent. However, even with an expected economic growth rate equal to the period between 1973 and 2003 (average 1.68 per cent for the United States, Canada and Europe, and 2.47 per cent for the rest of the world), Western states' share of global GDP is expected to sink below 30 per cent by 2050 – less than it was in 1820 (Goldstone 2010). Some 80 per cent of future economic growth will take place outside the Western world. This will, in turn, result in a massive transfer of markets to other regions, in which the middle class population is booming. According to World Bank projections, the middle class in these parts of the world will reach some 1.2 billion people by 2030. This number is higher than the expected number of people in Europe, Japan and the US put together (Goldstone 2010). The trends inevitably point towards an already ongoing transfer of economic power away from the Western world towards emerging powers, such as Brazil, Russia, India and China (the so-called BRICs), but also towards states like Indonesia, Mexico and Turkey. Views differ as to the sustainability of economic growth in the BRICs in particular – Russia, for example, should perhaps rather be seen as a declining power. Economic forecasts are, in any case, uncertain, and black swans will occur, as seen in the ongoing global financial crisis.

However, the shape of things to come is already evident in massive cuts in national defence budgets. For example, in its military strategy from 2011, the US Joint Chiefs of Staff regard US foreign debt levels to be a significant threat to national security (US Government 2011). US foreign debt already amounts to more than US$3.5 trillion. In the future, the US will be spending more than 7 per cent of its GDP just to pay the interest (US Government 2010, p. 21). In comparison, the US national defence budget equalled only 4.7 per cent of its GDP in 2010. President Obama has signalled that the US will have to cut defence expenditure by US$78 billion over the next couple of years, but others expect even tougher cuts (see e.g. Mandelbaum 2010). This will inevitably lead to a change in priorities regarding where, when and how the US will use military force in the future. These changes can already be observed in a general shift of military focus away from the relatively stable Europe towards the Asia–Pacific region. However, future cuts will also affect what is often referred to as the US' 'hidden export', i.e. the securing of global trade and communication routes – or the so-called global commons – and the peace and stability ensured by the continuing US military presence in several regions and conflict zones around the world (US Government 2010, p. 22).

In Europe, we may also observe a general trend towards a tightening of defence budgets. Given the economic realities of the ongoing crisis of the Eurozone, this trend will only grow stronger in the years to come (see US Government 2008, p. 32). Yet it is worth noting the fact that together European states still spend more than US$200 billion per year – more than double the amount of China – and still have some two million soldiers on active duty (see International Institute for Strategic Studies 2008, chapter 4).[3] The problem is that only a limited share of these forces is deployable internationally, while effective spending is hampered by unnecessary (and necessary) duplication of capabilities. The solution, as held by most security and defence policy analysts today, lies in more integration by way of 'pooling and sharing' of capabilities, and more role specialisation amongst the European states. In light of the financial crisis there seems to be a growing political momentum for more military collaboration, as reflected in the NATO initiative on 'smart defence' (see NATO 2010) and the EU's Ghent framework for pooling and sharing of military capabilities (Germany/Sweden 2010). However, as remarked by Tomas Valasek (2011): 'What makes obvious sense to experts and officials looks very different to national defence ministers.' Collaboration takes years to yield rewards and may initially cost more than it saves, while also carrying real political risks, insofar as opposition politicians and journalists will often accuse defence ministers of undermining national sovereignty. Europe knows it will have to spend smarter, but it is a slow train coming.

Until the incentives become strong enough for policymakers to move beyond the politically safer route of inaction, therefore, Western states will have to explore other options that may save money in the short to medium term. Privatisation is but one obvious solution in this regard. Private Military Companies (PMCs) have previously played an important role in the conflicts in the Balkans in the 1990s and in numerous civil wars and armed conflicts in countries such as Liberia, Sierra Leone and the DRC. The US-led intervention in Iraq in 2003 is often described as a breakthrough for the ongoing privatisation of military services in Western states. And the tendency is increasing, despite serious issues raised with regard to, for example, what used to be Blackwater's role and activities in Iraq (see e.g. Avant 2004, 2005). This trend may, on the one hand, be reversed given greater awareness of the ethical issues and command and control challenges incurred by outsourcing military force. On the other hand, it may also grow stronger with the increased pressure on national defence budgets. Policymakers may, perhaps in international military operations in particular, be tempted to let increased competition in a growing marketplace for military services push the costs of intervention down. By outsourcing tasks to PMCs, policymakers may also avoid the political costs of having the nation's young men and women return in body bags. Such use of 'foreign legionnaires' or 'colonial forces' is, of course, a practice that has been around in states like the UK, France and Spain for centuries.

Another option may be to rely more on smaller deployments, often using Special Forces to carry out limited raids, or to make use of mentoring and training of local forces to avoid having to bear the costs of deploying and sustaining large units in the field. NATO has highlighted the increased use of 'indigenous forces' as a key component in its future international operations (NATO ACT 2009). It is held that having the host nation carry out stabilisation may even increase the operational effect in theatre, while avoiding the problem of mission creep. However, such operations necessarily involve large cadres of Western officers who have the language and cultural skills to carry out mentoring and liaison tasks in often extremely hostile environments. These and other challenges are discussed by Guro Lien in Chapter 5, while the increased use of Special Forces as a key Western asset in international military operations is treated in depth by Iver Johansen in Chapter 4.

Technological trends – back to basics

Most future studies hold that scientific and technological developments will grow ever faster and spread to new areas. A halt in or reversal of technological developments is not considered likely, even if there are historical examples of technology and knowledge lost.[4] Exactly *what* the future will bring of technological innovation is impossible to anticipate, but much is expected to happen in the areas of information technology, nanotechnology and biotechnology. In addition, new fields of technology may emerge. These developments will open up new opportunities for growing numbers of people. But they will also lead to new challenges.

First, society today is becoming more reliant on information and communication technology, but the infrastructure for these technologies is vulnerable. NATO's strategic concept cites cyber attacks as a key threat against the Alliance in the years to come (NATO 2010, para. 12). Second, increased access to communication technology may destabilise countries, as seen, for example, in the demonstrations that led to the downfall of the regimes in Tunis and Egypt, in which mobile phones were used actively to spread information amongst the demonstrators. The ongoing Arab Spring continues to demonstrate the difficulties totalitarian regimes face in trying to control their populations in today's information age. Third, increased awareness of inequalities within and between countries and regions, as a result of increased access to cheap communication technology, may lead to anger and frustration amongst the have-nots. Fourth, new knowledge and technology may be utilised by potential adversaries, referred to below as terrorists, to harm societies in ever new ways. The result of these developments may be local and regional conflicts that may require international intervention. The dynamics within these eventual operations may, in turn, change as the local populations' expectations of and attitude towards the intervening force will be affected by greater awareness of the outside world, instant knowledge of negative events on the ground, and the ability to mobilise and coordinate large crowds of people as mobile phones and social media become effective means of 'command and control'.

Technology and innovation have played a particularly important role for the military. In his essay 'The Forgotten Dimensions of Strategy', Sir Michael Howard shows how technological superiority has had a decisive effect upon the outcome on almost every war in (modern) history (Howard 1983, p. 104). Yet Howard cautions against forgetting the social, logistical and operational dimensions of strategy. Throughout military history, the importance placed on technological and human factors has tended to shift as technological innovations have been met with effective counter measures. However, a brief look at the last 20 years shows that the faith in advanced technology as *the* solution to all challenges of war has gained an increasingly dominant role, as epitomised by the so-called 'Revolution in Military Affairs' (RMA). This is again changing.

In the US, the technological dimension is now being toned down (see e.g. Department of the Army 2009). That is, the continued emphasis on and investments in technology will not go away, but it will not necessarily claim as central a role as it has in the past. These developments will have to be seen in conjunction with the economic challenges discussed above. For example, in the US there have been cuts in the F-35 joint strike fighter programme, while prestige projects such as the Future Combat System have been cancelled. The shift of focus is evident in the US Army's capstone doctrine from 2009, which states that RMA proponents have disregarded many of the universal characteristics of war, while at the same time putting too much emphasis on new technologies and capabilities. It is held that concepts emphasising long-range precision-guided weapons in particular have disregarded war's political, cultural and psychological context (Department of the Army 2009, p. 6). The resonance with Howard's 20-year-old warning is obvious.

Whether the emphasis will move back to technology in the future is impossible to say. However, it is worth reflecting on the fact that future technological innovation may not necessarily take place inside the Western defence industrial complex, which so far has given birth to innovations such as the Global Positioning System (GPS), the Internet, etc. Major innovations today are developed by civilian companies, such as Google and Apple. These companies are still predominantly located in the Western world, which continues to score higher on innovation indexes than countries such as China and India. But the gap is expected to be reduced over the next ten years, as these countries learn to utilise their comparative advantage in new fields of technology (US Government 2008, p. 13). In any case, in the years to come it is expected that the technological superiority, which has benefitted Western military forces for decades, will gradually wither, as '[f]uture adversaries will use commercial off-the-shelf capabilities (to include information technology) to construct a well-organised, dispersed force capable of complex operations' (Department of the Army 2009, p. 13). Understanding how people might use technology will be more important than the technology itself.

At the core of many of the emerging challenges described so far is the simultaneous vulnerability that increased technological dependency creates. This is a challenge for Western military forces in particular. Most weapon systems today are to some degree reliant on network technology and/or GPS. Neutralising these systems, for example through various forms of jamming, will have major consequences for how Western forces operate. The US Army's *Capstone Concept* takes as a point of departure that, in the future, US forces cannot expect to be able to operate with all systems intact. Protection of one's own, and the ability strike against an opponent's, networks and systems – i.e. cyber warfare – will, therefore, grow in importance. This is reflected, for example, by NATO's signalling that a cyber attack may trigger a collective response under article five of the North Atlantic Treaty (NATO 2010). Insofar as the frequency and effects of cyber attacks increase significantly, or for example effective GPS jammers become commercially available, it may trigger a trend towards more robust, low-tech solutions. The growing importance of the cyber domain for international military operations is covered in depth by Siw Tynes Johnsen in Chapter 9.

Finally, technological developments will affect how Western forces operate on the battlefield whether at home or abroad. The pressure on European forces in particular to transform and adapt to stay interoperable with US forces will continue to set the agenda for some time to come. But, in the long run, an evening out of the technological differences between the current haves and the have-nots may also have a democratising effect on how multinational operations are carried out, as emerging actors are challenging the traditional lead nations.

New conflicts, new actors, new missions

The trends discussed above will give rise to new types of conflict, new challenges and solutions within existing types of missions, and also completely new types of missions and approaches, all as new state and non-state actors are entering the scene. The Arab Spring illustrates, for example, how technological developments, such as a world in which 'everyone' has access to cheap information technology, fuel the disappointment young people especially feel, as their expectations of democratic development and increased living standards are not met – or are not met soon enough. It is becoming increasingly difficult for even authoritarian regimes to control their own citizens. Mobile phones as a tool for mass mobilisation and instant spread of information have proved to be very effective weapons. Moreover, increased awareness of inequalities within and between countries and regions may also cause conflicts to spread, as seen in waves of demonstrations in states that are subject to rapid development, such as Turkey and Brazil.

Limited access to natural resources also carries a significant conflict potential, not least in light of the economic and demographic trends described above. Limited access to water in the Middle East and Africa may spur new water wars. The consequence may be a demand for international action to alleviate humanitarian crises or intervene in conflicts in operational environments that will offer severe logistical challenges for international forces. Finally, competition for coal, steel and other minerals, on which industrialising states especially depend, may also lead to new conflicts in the future.

In other words, there is little to suggest that the *demand* for international military operations will go down in the future, even if this demand is not necessarily as hard felt among the states at the receiving end of the interventions, nor among some of the emerging powers such as China, India or Brazil. The key problem is rather to be found on the *supply* side, as the capacity and will among the states that have traditionally championed international interventions in the past is bound to wither. This is not to say that one always has a choice of intervention, as is sometimes implied by distinguishing between 'wars of choice' and 'wars of necessity'. Insofar as national interests in a globalising world will be increasingly threatened outside national borders, international military operations far away from home may rather become 'wars of necessity' in the future. The controversies surrounding these operations will, at the same time, increase as their legitimacy comes under pressure, domestic debate grows and the more altruistic motives behind these operations are toned down.

However, this development must be seen in relation to another trend, which has partly moved in the opposite direction, and which follows much in line with Sir Rupert Smith's thoughts on war among the people. Protecting *people* regardless of nationality has over the last decade become a key priority for and in operations across the entire conflict spectrum. In light of previous failures, the UN in particular has made protection of civilians a top priority in its peacekeeping operations today. The adoption of the principle of the 'Responsibility to Protect' (R2P) represents the latest attempt to address the issue of mass violence against civilians happening in the first place. However, as Alexander W. Beadle discusses in Chapter 11, protecting civilians has also become an objective beyond the moral imperative. Protecting people has become a military–strategic necessity in most types of operations, including operations where it is not the primary objective; this first and foremost derives from the changing nature of contemporary conflict in which civilians are increasingly targeted. Yet a key question is whether emerging powers, such as China, Russia, India and Brazil, which have all grown critical of using armed force in instances where regimes attack their own populations, will support these types of counter-regime operations in the future. Such operations have also posed a number of challenges to the intervening parties in the recent past, which raises questions about their desirability with those who have championed them in the UN, an issue that is discussed by General (Ret.) Sverre Diesen in Chapter 3. As we shall see, counter-regime operations may still be hard to avoid.

In any case, a rather obvious consequence of the trends discussed in this chapter is that, as a result of economic expansion and a general transfer of power from the Western world to emerging powers to the South and East, new actors will (have to) take on an expanding role in and for international military operations in the time to come. As an example of the inherent logic of these developments, take China's current role and interests in Africa: currently, China has more than 100,000 'settlers' in Angola – young, highly educated individuals, who have contributed to rebuilding critical infrastructure, such as roads, and restarting the economy after decades of civil war. Even if the image of China's role in Africa is complex – China's interests in South Sudan are, for example, seen as more controversial – the Chinese enjoy much legitimacy and respect in many African countries. This presence leaves a very different impression of China than the usual threat images painted in the West (see e.g. Brautigam 2009). Nevertheless, there is little reason to doubt that China will move to protect its interests if, for example, threatened by civil war. The international community will then have to decide whether an eventual Chinese operation should be sanctioned by the UN, who would lead it, whether to participate in the operation, etc. These questions will, in turn, have wider consequences for the UN as the central sanctioning authority in matters concerning war. China is already the largest troop contributor to UN missions amongst the permanent members of the Security Council, and is, therefore, an actor with which Western forces will have to cooperate to stay relevant as global security providers in the future. Generally, Western forces will have to prepare for more cooperation with states like Brazil, India, Indonesia and others – with the operational and cultural challenges this will pose.

These expected changes in global security provision must also be seen in relation to the increased awareness of challenges to the so-called 'global commons': the high seas, the atmosphere, outer space and cyberspace. Western states, with the US at the forefront, have for long secured free access to and passage through the commons through international regimes with various degrees of formalisation, such as the Law of the Seas. Whether these regimes will be respected by emerging powers remains an open question, especially if they are seen to run counter to shifting national interests. The US' traditional dominance in the world is under pressure, which may lead to more of an international division of labour when it comes to keeping the commons open. The naval operations to counter the threat of piracy in the Gulf of Aden represent but one example of an international effort to secure free passage in one of the commons. Similar types of international military operations may become more common in the future. Such operations may involve 'taking out' hackers operating in states or cities with limited governmental control, taking out servers, taking out pirate bases onshore, supporting other states' cyber warfare, etc. A common denominator for these operations is that they will be limited in scope and duration. This will, in turn, increase the demand for highly specialised assets capable of operating alongside other states' forces and/or units from other branches within and beyond the military.

In sum, the trends described above may point towards more limited, interest-based operations, following a period in which humanitarian operations – or at least an emphasis on humanitarian motives for intervention – have been the norm. This may require some form of role specialisation in and between those states that want to take part in these operations in the future. Issues such as the legality and legitimacy of intervention are also becoming more muddled, as seen for example in changes in the way that consent is handled and mandates are given. These are issues that are, in turn, complicated by the fact that many challenges – and thus solutions – today are transnational. Economic shifts suggest that Western countries have to find cheaper forms of intervention, such as Military Advising and Assistance operations or privatising military tasks, a trend that has been coming for quite some time already. There is still the danger that the global community will stumble into operations of the kind that they do not want to carry out, typically involving state or nation building, tasks that require enormous resources and not least time. The recurring question is still how to devise appropriate exit strategies, not necessarily how to avoid these operations. And then we have the new forms of conflict, or old conflicts in new guise, including tomorrow's urban conflicts.

Nine mission types for twenty-first century international intervention

In this book, we try to cover some of the emerging challenges drawn from the analysis of key global trends and trajectories – and seek to identify the utility of force in dealing with them – by analysing nine carefully selected mission types, which we believe will be relevant for future international military operations. We arrived at this selection by considering two ways that new challenges for the international use of military force might arise. The first is how present mission types will evolve in a conflict environment shaped by global trends. The following belong in this category:

- high-intensity operations (Chapter 3)
- counter-insurgency operations (Chapter 4)
- military advising and assistance operations (Chapter 5)
- special forces operations (Chapter 6)
- UN operations (Chapter 7).

The second way is how the global trends may give rise to completely new mission types. The following belong in this category:

- new urban operations (Chapter 8)
- transnational operations (Chapter 9)
- cyber operations (Chapter 10)
- protection of civilians in operations (Chapter 11).

We could have chosen other mission types, or other ways of grouping them. Some would, for example, question why we have chosen to include a chapter on UN operations, which is not exactly a new type of mission. But we would argue that the UN is such an important actor in international military operations, and that it imposes such essential boundaries on other actors' behaviour in all international interventions, that it deserves to be analysed as a mission type of its own. This is not least because changes to the way the UN operates, and its role as the key sanctioning authority in matters relating to war, will have huge implications for international interventions in the future. In fact, all of the first five mission types analysed are partly about salvaging some of the hard-won lessons of previous operations, rather than succumbing to the politically tempting, but arguably false, idea that they can be avoided in the future. In the chapters on military advising and assistance and special operations, we also explore the feasibility of carrying out these operations more cheaply and efficiently in the future. Some of the mission types include elements of nation building, which some might argue deserved its own chapter. However, we have chosen to deal with this challenge by looking at nation building as intrinsic to many of the mission types and by drawing some general observations across the mission types in the concluding chapter.

Others will miss chapters on, for example, anti-piracy operations and littoral operations, which have received some attention lately (see e.g. Kilcullen 2013). However, we have deliberately avoided purely naval operations, retaining rather a focus on land operations. Littoral operations, in turn, are touched upon in the chapter on the new urban operations, as the trends towards littoralisation can be seen as more or less intrinsic to urbanisation. Although other mission types may also be missed by some readers, it is necessary to stress again that for foresight it is sufficient that a scenario is plausible. The sample of scenarios or mission types does not need to be exhaustive. It only needs to find sufficient interesting new challenges to be helpful. We believe that the nine mission types selected for this volume fulfil that objective.

Finally, our instructions to each author were that they analyse their mission type using five generic parameters, which are set out in Chapter 2, while taking an explicitly forward-looking perspective. But, to encourage originality and out-of-the-box thinking, we gave them some freedom as to how to approach each mission type, and how to relate to the parameters in their analysis. Accordingly, some of the chapters take a case study approach, some use a more traditional scenario approach, while others take a more general approach. The result, we believe, is a truly innovative perspective on international military operations in the twenty-first century, which will prove helpful to practitioners and policymakers, and thought provoking for anyone interested in issues of international intervention today.

Notes

1 The problem is far from new. Philosophers such as David Hume, John Stuart Mill and Karl Popper in particular were all preoccupied with the problem of induction in logical thinking – i.e. drawing general conclusions from the specific, and making future predictions based on the past.
2 Cognitive closure is a term borrowed from psychology, but it is often used to explain why intelligence analysts and decision makers in general have difficulties breaking out of established thought patterns.
3 According to the Stockholm International Peace Research Institute (SIPRI) Military Expenditure Database, China spent US$99.8 billion on defence in 2009, although the real number is probably higher.

4 The Mayans performed brain surgery, and the early Middle Ages were recognised by a generally lower level of development than previous times. A more recent example can be found in Germany's decision to abandon nuclear power due to its potential negative consequences.

References

Avant, D. (2004). "The Privatization of Security and Change in the Control of Force". *International Studies Perspectives* 5(2), 153–157.

Avant, D. (2005). *The Market for Force.* Cambridge: Cambridge University Press.

Barnett, T. P. M. (2004). *The Pentagon's New Map: War and Peace in the Twenty-First Century.* New York: Putnam.

Brautigam, D. (2009). *The Dragon's Gift: The Real Story of China in Africa.* Oxford: Oxford University Press.

Daniel, D. C. F., P. Taft and S. Wiharta, Eds. (2008). *Peace Operations: Trends, Progress, and Prospects.* Washington, DC: Georgetown University Press.

Department of the Army (2009). *The Army Capstone Concept.* Department of the Army Headquarters, US Army Training and Doctrine Command, TRADOC Pam 525–3–0.

Gardner, D. (2010). *Future Babble: Why Expert Predictions Are Next to Worthless, and You Can Do Better.* Toronto: McClelland and Stewart.

Germany/Sweden (2010). *Pooling and sharing, German–Swedish initiative.* Berlin and Stockholm: Food for thought paper. 10 February.

Glenn, J. C. and T. J. Gordon, Eds. (2009). *Futures Research Methodology Version 3.0.* Washington, DC: AC/UNU Millennium Project.

Goldstone, J. A. (2010). "The New Population Bomb". *Foreign Affairs* 89(1), 31–43.

Gow, J. (2006). "The New Clausewitz? War, Force, Art and Utility – Rupert Smith on 21st Century Strategy, Operations and Tactics in a Comprehensive Context". *The Journal of Strategic Studies* 29(6), 1151–1170.

Gray, C. (2005). *Another Bloody Century: Future Warfare.* London: Weidenfeld & Nicolson.

Haug, K. E. and O. J. Maao, Eds. (2011). *Conceptualising Modern War.* London: Hurst & Co.

Hegre, H., H. M. Nygård, H. Strand, H. Urdal and J. Karlsen (2013). "Predicting Armed Conflict, 2010–2050". *International Studies Quarterly* 55(2), 1–21.

Howard, M. (1983). "The Forgotten Dimensions of Strategy". In *The Causes of Wars.* M. Howard. London: Temple Smith, 101–115.

Howe, N. and R. Jackson (2008). *The Graying of the Great Powers: Demography and Geopolitics in the 21st Century.* Washington, DC: The Center for Strategic International Studies.

International Institute for Strategic Studies (2008). *European Military Capabilities: Building Armed Forces for Modern Operations.* London: Institute for Strategic Studies, Strategic Dossier.

Kilcullen, D. (2013). *Out of the Mountains: The Coming Age of the Urban Guerrilla.* New York: Oxford University Press.

Mandelbaum, M. (2010). *The Frugal Superpower: America's global leadership in a cash-strapped era.* New York: Public Affairs.

Mayall, J. and R. S. de Oliveira, Eds. (2011). *The New Protectorates: International Tutelage and the Making of Liberal States.* New York: Columbia University Press.

NATO ACT (2009). *Multiple Futures Projects – Navigating towards 2030 – Findings and Recommendations*. Norfolk Va: NATO Allied Command Transformation.

NATO (2010). *Strategic Concept for the Defence and Security of the Members of the North Atlantic Treaty Organisation*. Adopted by the Heads of State and Government, Lisbon, 19 November.

Norheim-Martinsen, P. M., S. Glærum and H. Fridheim (2011). *Planning for the Future – A Norwegian Perspective*. Report to NATO specialist meeting, SAS-088, 'Long Range Forecasting of the Future Security Environment', Stockholm, 11–12 April.

Paris, R. and T. D. S. Sisk, Eds. (2009). *The Dilemmas of Statebuilding – Confronting the Dilemmas of Postwar Peace Operations*. London: Routledge.

Smith, R. (2005). *The Utility of Force: the Art of War in the Modern World*. London: Penguin Books.

Strachan, H. and S. Scheipers, Eds. (2011). *The Changing Character of War*. Oxford: Oxford University Press.

Taleb, N. N. (2007). *The Black Swan: The Impact of the Highly Improbable*. London: Penguin Books.

UN-HABITAT (2008). *State of the World's Cities 2008/2009: Harmonious Cities*. New York: United Nations Human Settlements Programme.

US Government (2008). *Global Trends 2025: A Transformed World*. Washington DC: National Intelligence Council.

US Government (2010). *The Joint Operating Environment*. Norfolk, Va: United States Joint Forces Command.

US Government (2011). *The National Military Strategy of the United States of America. Redefining America's Military Leadership*. Washington DC: US Joint Chiefs of Staff.

Valasek, T. (2011). *Governments Need Incentives to Pool and Share Militaries*. London: Centre for European Reform.

Vautravers, A. (2010). "Military Operations in Urban Areas". *International Review of the Red Cross* 92(878), 437–452.

Webster, D. M. and A. W. Kruglanski (1997). "Cognitive and Social Consequences of the Need for Cognitive Closure". *European Review of Social Psychology* 8(1), 133–173.

2 Five parameters for analysing international military operations

Tore Nyhamar

In the first pages of the previous chapter, the overarching questions of this book were set forth. The rest of Chapter 1 discussed the first question: how will trends in demography, economics and technology shape future operations? This chapter now turns to the question of what future military operations may look like in terms of tasks and mission types. As set out in Chapter 1, the book discusses nine mission types: high-intensity operations; counter-insurgency; military advising and assistance operations; Special Operation Forces; United Nations operations; urban operations; transnational operations; cyber operations; and operations to protect civilians.

Based on these nine mission types, we seek answers to the second overarching question of this book: what will be the utility of force of the future operational demands on international military forces in different missions? This chapter tries to assist in this endeavour by accomplishing two things. The first is to provide a generic framework to describe military mission types based on five parameters that can be used to define future military operations. The second purpose of the chapter is to offer a broader perspective on important evolving and future trends than is possible in the analysis of one specific mission or operation type. The chapter uncovers and describes possible future changes within each parameter.

Let us then move to the role of the parameters in shaping the next nine chapters dedicated to different specific mission types. The chapter describes five generic parameters, which are to be used to discuss future international military operations: mandate; consent; conflict intensity; operational environment; and relative force composition and strength. These parameters single out the general characteristics of an international operation determining the operational requirements of the international force. Combined, the parameters will shape the tasks of an international force in an operation. Moreover, they provide key questions that structure the chapters on the different operations. They are not a straightjacket; the individual chapters on operations do not weigh or rank the parameters. On the contrary, the authors have been encouraged to provide a narrative of the trends in their mission type. What the chapters do, when appropriate, is to reflect upon these five parameters in the context of each mission type. The term parameter is used in its common meaning: to identify a characteristic or a feature to define a particular system. Parameters hone in on the important in describing the operational challenges of the operational context in any international military operation. The description of each parameter will define what it is, spelling out the most striking recent developments affecting operations. There follows an assessment of the parameter's past values, its present status and, when possible, future trends and trajectories.

Mandate – from bystander to agent for change

The mandate matters because it determines the objectives of the operation. However, we have deliberately left out a detailed discussion of the political objectives for the operation because our objective is to be prepared for the future rather than for what the next operation might look like. In broad and general terms, many of the chapters suppose that stability is a direct or indirect goal to be promoted. The goals of the states taking part in an international operation will vary, and the complexity will increase further if domestic political opinion is included. There might also be variation across mission types, adding a further layer of complexity. When appropriate, the chapters may delve into political goals specific to the mission type. Today an international force may be mandated to reconstruct a collapsed government at the same time as it fights insurgents. Undertaking such tasks began in the early 1990s when there was hope that international forces could contribute to a new world order. Today, chastised by setbacks in Rwanda and Bosnia, emotions are mixed, ranging from intervention fatigue in the wake of operations in Afghanistan, on the one hand, and optimism caused by the progress of R2P on the other.

Most international military operations have an explicit mandate from the UN Security Council that, in broad terms, determines what the mission ought to accomplish, and authorises the means to do so. The task of the international force is to find ways to deploy military personnel to reach the objectives set out in the mandate. One distinction must be made between operations carried out by the UN itself – the blue helmets – and operations carried out by a regional organisation with a mandate from the UN. The current operation in DRC, *the United Nations Organization Stabilization Mission in the Democratic Republic of the Congo* (MONUSCO), is a UN operation, whereas Operation Unified Protector (OUP) was an operation run by NATO to carry out the objectives of resolution1973 (2011) of the UN Security Council from 17 March 2011. International interventions without a Security Council mandate are rare. The United Kingdom's intervention in Sierra Leone in 2001 and NATO's Kosovo campaign in 1999 are notable exceptions.

Historically, the mandate presupposed the signing of a peace agreement between two well-defined entities (states) in the past, typically leading to a UN peacekeeping operation. UN peacekeeping was founded on the principles of neutrality, impartiality, consent and the non-use of force. These principles constituted a related whole (Bellamy *et al.* 2004, p. 308). They were inextricably linked, because if one principle was altered, the others were affected. The history of UN mandates after the cold war is largely the history of coming to grips with the effects – intended or unintended – that changing one principle will have on the others within this related whole. In particular, three new trends in mandates have challenged the cold war principles: authorising international forces to use more force; giving them more ambitious tasks; and tasking them with the protection of civilians.

First, the UN mandate in the 1991 Gulf War, for example, authorised the use of force to redress the occupation of Kuwait and was directed against Iraq. As it was triggered by a clear breach of international law, an armed attack on a sovereign state, the Gulf War mandate only represented a minor change in international practice. Today, many UN mandates authorise 'all necessary means' to achieve the objectives of the mandate (i.e. they are authorised under Chapter VII of the UN Charter). The change this constitutes, however, only becomes apparent in the context of inclusion of new objectives in today's mandates.

Second, in 1989–1990, the operation in Namibia, *United Nations Transition Group (UNTAG)*, was the first with a broad political mandate to organise free elections and supervise the establishment of a new state (Bellamy *et al.* 2004). In addition, UNTAG also certified the process. When the international force is tasked with implementing state building in this way, there is a strong possibility that some group will oppose the broad political goals or the way in which they are implemented, confronting the international force with the question of whether to use force or not.

Third, since the 1990s, the task of the protection of civilians is an emerging trend in the mandates of international operations. In 2005, the majority of states in the UN General Assembly voted in favour of the R2P. It has taken some time for the principle to yield practical consequences, but the implementation of UN Security Council Resolution 1973 (2011) on Libya and the following operations meant that the use of force was for the first time justified by protection of civilians alone.

In a discussion of the current status of impartiality in UN operations, Jane Boulden (2005, p. 150) makes the useful distinction between mandates and implementation of mandates. The implementation process has become separate from the mandate for three reasons. The first is that it is undertaken by different people. The diplomats in the Security Council shuffling back and forth to craft an agreement to intervene are other than the soldiers and civilians who are deployed to implement the agreement. The second is that in today's operations the implementation process itself may undermine impartiality, the agreement or the mandate (Boulden 2005, p. 153). For example, the losing party in the first election may be genuinely committed to the political process but lose faith in the agreement due to widespread electoral fraud. Finally, it allows for the possibility that force may be used, and used more against some parties on the ground than others, and the operation still be considered impartial because the actions are occurring in fulfilment of an impartial mandate. This is the Western position on what happened in fulfilment of Security Council Resolution 1973. The Resolution imposed a weapons embargo and authorised the use of 'all necessary means' in order to protect civilians, the operation's objective (UN 2011). The subsequent debate between the Western powers on one side, and Russia and China on the other, illustrates the potential problems in agreeing the actions necessary to fulfil a particular mandate. The novelty is that, in today's operations in which the intervening force may take sides, actions in implementation that are in themselves not impartial may be justifiable in terms of the mandate.

Yet even if the impartiality of a mandate and its implementation are inextricably linked, they are not identical. First, if the mandate is not impartial, the implementation of it cannot be impartial. More generally, the mandate *may* matter because, if it is flawed, the force tasked with implementing it may essentially have an impossible mission. Second, poor implementation may in itself be sufficient to lead an otherwise sound mandate to fail. The international force needs flexibility in implementing the mandate but this will often also leave it with decisions that are political. Ideally, these should be settled by their own political leaders but in modern operations they are often left to the military (Ricks 2009, pp. 154, 164). How to improve the dialogue between desirable political goals and what is achievable on the ground is not clear, but it is clearly necessary to move beyond the separation of political goals and military means so typical of Western militaries (Schadlow and Laquement 2009, p. 114; Simpson 2012, p. 14). If mandates are not impartial, not everyone on the ground is likely to consent.

Consent – from international legality to recipient legitimacy

To consent is to accept or approve of what the international forces are there to do. It obviously facilitates the task for the force if everyone approves of the objectives of the operation. However, today's increased ambitions have made the international force an actor in the conflict. Managing the strategic, operational and tactical consent of local actors, and the interplay between these levels, is a new challenge that now may confront an international force.

In the past consent was the bedrock principle of UN operations and mandates. In many ways, it still is. In the era of traditional peacekeeping during the cold war, consent was largely about the legality of the mandate and the process that produced it because international interventions were an exception to the non-intervention principle. Therefore, in cold war peacekeeping, consent was determined by the legality of the intervention prior to the deployment of the force, and impartiality ensured that the use of force was redundant in the mandate.

At present, the increased role of international forces in shaping domestic politics in the recipient countries has, in some ways, undermined the principles of traditional peacekeeping. As discussed above, consent, impartiality and the non-use of force constituted a related whole (Doyle and Sambanis 2006, p. 308). An increased role in local politics tends to make detached impartiality difficult for the international force. Today, once the operation is underway, the actors in the area of operations matter more. There are two main groups that may give their consent. The first is located at the national political level, where the attitude of the government figures prominently. The second main group is the people, encompassing what may be called public opinion. The degree of consent to what the international force is doing may vary amongst different groups and over time. Consent from the government of the host nation is always an important, and sometimes a necessary, condition to embarking upon a military mission. Political consent, however, is far from simple and often domestically contested (NYU Centre 2011, p. 12).

When the objective is to alter local politics, the consent of new groups becomes vitally important for the operation's prospects of success. Moreover, consent becomes more of an issue of legitimacy than legality. It is not sufficient that the intervention is legal by being firmly rooted in international law; it must be *legitimate* in the eyes both of political elites and of various segments of the population in the area of operations. In modern operations, the consent of various layers of the population has thus become critically important; indeed according to Smith (2007, pp. 269ff.) failing to understand *the* defining characteristic of war among the people renders modern military operations without utility. First, consent has become a more complex concept. It used to be about avoiding the great powers changing policy. Now it is about support from different layers of the population in a foreign country. Second, consent has become a dynamic, malleable concept. Consent is something that the international force can be called upon to manage. In modern operations, the troops may need to consider consent from many actors simultaneously. The concept of consent has become a key concern to *manage* for the leadership of the operation.

Moreover, the need the international force has for consent if it is to succeed also varies with the type of operation and its corresponding objectives. In a military operation to unseat a regime, there is obviously little hope that it will consent to step down. On the other hand, it is difficult imagining an international force contributing to building governing capacity without consent from that government to accept assistance.

Three important elements of an international operation that domestic local actors may or may not consent to are: the deployment of an international force; the political process that produced the agreement that the international force is there to implement; and the mandate of the force. Consent to these three elements is interrelated and mutually reinforcing but not identical. Even failed states usually have a government that may give its consent to the entry of an international force. The problem is that this has little relevance because the government has little influence over actors and events in the deployment area. There, the consent of different actors may be critical to the force's success. In such situations, the management of consent by whom to what becomes critically important. States will usually be reluctant to commit troops to an area in which there are influential actors who do not want them there. Both international law and common sense suggest that the consent of the population and key political players is desirable and indeed necessary.

If the international force deploys in support of a political process, it is no trivial matter to judge its prospects for success, whatever that is in the particular case. In Iraq, the situation was dire in 2006, but the regime in Baghdad was able to pull back from the brink of collapse. In Afghanistan, on the other hand, the situation looked promising in 2002, but looks decisively more troubled at the time of writing. Finally, the local government may be willing to consent to political compromise, but if the international force has far-reaching political reform as its mandate, formally or informally, the local government may object to this. Both in Afghanistan and Iraq, the international forces lowered their ambitions when consent from key players and groups of the population was withheld.

The attitude of the population matters. The question of whether a peace accord is best interpreted as an agreement between the belligerent parties (as it usually is) or an agreement with the population still remains. An intervention force must find a way to handle local power holders, as it is a necessary but not sufficient requirement to accomplish the mission. First, local power holders are influenced by the attitudes of the population. Second, the attitude of the population may be all there is an area in which all formal authority has collapsed. Clearly, the degree of the consent of the population is not a constant factor. South African forces reported that they were viewed positively in Burundi and DRC, where they succeeded in providing basic security and access to basic needs, and in a less friendly way in Sudan, where they failed to do so (Heinecken and Ferreira 2012, p. 44). To increase the consent of the population is an explicit objective in counter-insurgency operations, and an implicit objective in most other operations.

The management of consent is further complicated by a logic that Doyle and Sambanis (2006) call the 'obsolescing welcome'. The position of an international force and the UN is never stronger than at the moment of deployment. Then the local actors are usually tired of a prolonged conflict and need peace and the international force and the institution most closely associated with the operation have not yet staked their reputation on the success of the operation. In short, the local actors need the forces of the operation more than the other way around (Doyle and Sambanis 2006). If the international force deploys, it typically moves through three phases. In the first, it enjoys a 'honeymoon period', a continuation of its strong position prior to deployment. The population adjusts to post-war reality and takes stock of the international force and its mission. In the second, segments of the population, often different segments from those that initially wished them welcome, decide about whether to support, embrace, tolerate or reject the presence of the international force. That decision is influenced by how the force provides security and how the credibility of managing the consent to restore sovereignty is managed. In the final stage, the intervention force faces the full weight of the duration dilemma and the obsolescing bargain when segments turn to various forms of obstruction and outright resistance of its mission (Edelstein 2010, p. 83). Managing consent during the various demands posed by this three-stage process is demanding.

What to do about warlords in the future clearly brings out some of the internal dilemmas involved. By definition, a warlord is an individual who control small slices of land, in defiance of genuine state sovereignty, through a combination of patronage and force (Marten 2006, 2011, p. 302). Since patronage is the ability to personally control the distribution of resources without oversight, usually excluding people who do not belong to the network of the warlord, the alienation of some groups is virtually ensured. Here the reliance on a personally controlled militia enters the equation. The UN Capstone Doctrine distinguishes between strategic and tactical consent (UN Department of Peacekeeping Operations 2008). A mission must gain and keep the consent of the key players in a conflict to succeed, but not necessarily of the local spoilers. Judging what is a local spoiler and what is strategic actor is a thorny question, but the analysis above suggests warlords tend to fall in the latter category. However, there are many local power brokers inside and outside of government that do not fall neatly into either category.

Drawing the line between spoilers and strategically important actors is further complicated by the issue of incremental obstruction (Johnstone 2011b, p. 14, 2011a, p. 177). An international force may face an opponent capable of slowly strangling the prospects of success through obstruction on many tactical issues without actually staging a direct frontal assault on the agreement. Slobodan Milosevic's actions in Bosnia from 1992 onwards are arguably an example. At what point does the tactic amount to a strategic withdrawal of consent?

Conflict intensity – from shooting to deciding about shooting

Conflict intensity determines what kind of experience the troops in the operation need to prepare for, and what kind of equipment they will need to achieve the objectives of the operation. The parameter has two dimensions: combat intensity and moral anguish. The former has moved from traditional peacekeeping at the low end of conflict intensity. Modern operations now include the entire conflict spectrum, often *within* the same operation. Modern operations, including so-called low-intensity conflicts, now include combat experience at the tactical level associated with industrial war, but are arguably more demanding as *varying* combat intensity seems to be the most taxing experience for the troops (Heinecken and Ferreira 2012, p. 41). Moreover, the fact that today's operations may take place among the people may make the troops face new kinds of psychological and moral anguish *in addition* to the traditional challenges associated with engaging in combat. Sherman (2010, p. 1) underlines that, based on testimonials of soldiers, the psychological anguish of war is also moral anguish. Soldiers from World Wars I and II clearly recognised that armies constituted bureaucracies that limited their choices, but still strove to retain personal accountability. This moral anguish experienced by troops can often be more acute in modern operations because troops are more likely to face morally challenging choices.

Clashes between militaries which are both organised in a way capable of integrating and coordinating different types of units used to be referred to as symmetrical conflict. It is symmetrical only in the sense that the forces are organised according to the same set of principles, as the purpose of all war is to achieve an asymmetric advantage. Clashes between militaries which are conventionally organised produce high intensity conflict. At the other extreme of the scale, in pure low-intensity conflicts, the international force will typically be engaged in non-military tasks like policing, supporting civil society and building local and national institutions. Here, opponents either do not exist or lack both the will and the means to carry out armed attacks.

To use military force is to apply violence (Huntington 1957, p. 11). Maximum violence is applied in conventional military battle, thus representing one extreme of the scale. Conventional military battle takes place between parties that both possess a large register of military capacities organised in different units supposed to work in an integrated and coordinated fashion towards a shared objective. For armies, examples of units would include infantry, artillery, armour, and engineers. In the land domain, operations involving integration and coordination between units having these different capacities are called 'combined arms operations'. Operations involving coordination between units from different services (army, navy and air force) are called 'joint operations'. In the clash between militaries consisting of units that are functionally different, and organised in ways rendering them capable of integrating and coordinating, maximum violence takes place. The term 'high-intensity warfare' captures the ability to produce this effect. The ability to sustain this effect results in industrial war (Smith 2007, pp. 107ff.)

All armed conflicts can be placed on a scale between high and low threat to own forces. Conflicts are rarely *either* extreme high intensity *or* low intensity, on the contrary most are a mix. One can simply treat each conflict as a specific mix of low and high conflict intensity, falling somewhere in between the extreme points of the scale. Today, where there are non-regular forces – militias or other armed groups – determined to resist an intervention force organised according to the principles of industrialised states, the result is low intensity conflict. It is above all the organisation aiming at integrating units with different capacities that set Western militaries apart and allow to them to dominate all others. The functional differentiation between different building blocks capable of acting together is also at the heart of what allows an army to militarily dominate militias or other armed groups. It is, however, not only the ability to carry out violence that influences conflict intensity. The *will* to commit violence matters. If the will is sufficiently strong, actors in the area of operations will use whatever means they do have. Opponents inferior in terms of military means will seek to compensate by using the means available to them, often in innovative ways. Examples of such unconventional means are suicide attacks, improvised explosive devices, kidnapping hostages. Moreover, opponents of the international force can simply attempt to create disorder, avoiding military confrontations as best they can (Smith 2007, p. 278).

The opponents will seek to operate in ways which favour their less sophisticated weaponry and organisation. They will find ways to create tactical situations where for a short period of time they are able to dominate a Western force with light assault rifles or rocket propelled grenades. A force organised according to Western military principles facing such an opponent has two alternatives. The first is to harness all its capabilities to unleash its full potential for violence on the opponents and their people. Historically, it is how most insurgencies have been defeated, but it is not an option for a Western military force, defeating the purpose of any modern military intervention (Smith 2007, p. 376).

The second is to use its conventionally organised units more as their opponents do (Arreguin-Toft 2006, pp. 34–47). For example, meeting the indirect approach of the opponent with our own indirect approach is at the heart of the idea of counter-insurgency (COIN). In practice, for the international force, it means that it must operate with smaller units, usually at platoon or troop levels, sometimes at the company, and only occasionally at the battalion level. The intervention force will still dominate militarily and have the ability to escalate beyond anything the opponents can do. How far the intervention force will dip into the full range of modern weapons systems will vary but remain below the conventional definition of low-intensity conflict. This definition made perfect sense in bringing out the differences between conventional battle in Europe and peacekeeping elsewhere. If considering the personal combat experiences of troops taking part in so-called low-intensity conflict, it makes less sense.

Their experiences will often be very different from those usually associated with low-intensity conflict. The small units in which the intervention force mainly operates may end up in firefights just as intense as those during conventional high-intensity conflict. This was what British and Danish troops experienced in Helmand in 2005–2007 (Marston 2008; Thruelsen 2009; Egnell 2011). This was what US troops experienced in Ameriya in Baghdad (Gentile 2008; Kuehl 2008). If intense fighting occurs sufficiently often to dominate the everyday operational experience of the troops, the idea that they are taking part in a low-intensity conflict seems odd. Moreover, an intervention force expecting to take part in a low-intensity conflict will come unprepared for the mission in terms of training and materiel and also mentally (Heinecken and Ferreira 2012, p. 48). In today's conflicts among the people, the battlefield penetrates on all fronts, forcing the soldier always to be battle-ready. Factors shaping the everyday experience include the duration and frequency of engaging in battle (albeit fighting in small units), the offensive/defensive approach of the opponent, the weapons involved, the rules of engagement, the experience of risk and danger, the length of pause between combat situations the troops have, how often they are on leave, whether it is safe to visit the market in the local village, the technologies available to make the presence of their own civilian life felt, etc. There are no clear front lines; Improvised Explosive Devices can be triggered by anyone in the crowd using mobile phone technology (Sherman 2010, p. 13).

But the actual battle experiences are not the only thing that has changed for modern troops. Modern technology – Internet and mobile phones – brings civilian life closer to soldiers when they are deployed. All soldiers also have a private, civilian life and in modern operations they are forced to switch between military and civilian roles more often. This contrast may in fact increase the intensity of the combat experience.

There is a need to move from the industrial war paradigm into modern operations in the realm of moral and ethics also. First, today's wars are wars of choice, making the moral justification less obvious. Second, soldiers fight in a different way in modern operations. The fight is usually carried out by small units with a higher degree of autonomy to decide how to fight than most units would have experienced in conventional high-intensity warfare in either World War. Today soldiers make more choices involving moral ambiguity, confronting ethical dilemmas more often. Third, modern operations may, to a varying degree, involve a complex mixing of tasks. Soldiers are asked to fight combatants, often dressed as civilians, and simultaneously mingle with civilians indistinguishable from them. Moreover, there may be no clear enemy only a plethora of actors motivated by their own concerns which may or may not lead them to take up arms against the international force (Simpson 2012, pp. 43–49) In addition to the psychological strain, this phenomenon may also confront individual soldiers with moral ambiguity and difficult decisions. Indeed, South African forces report that in addition to NGOs, local forces and the local population, just relating to a wide range of forces from different continents and countries was in itself challenging (Heinecken and Ferreira 2012, pp. 45–46).

Let us return to the question of the burden of the war soldiers fight, not just their individual conduct or the actions of the unit they belong to. The standard rebuttal of the feelings the individual soldier may harbour about the moral foundations of the war he or she is participating in is that the moral responsibility for that decision rests with others. Belonging to a hierarchical and coercive organisation, the individual soldier is exempted from responsibility for the justness of the cause of the war (Huntington 1957; Sherman 2010, p. 47). This traditional reasoning is based on conventional war, fought for the existence of the state or its national interest. Nevertheless, investigations after the mission seem to indicate that many soldiers still do retain responsibility, and that they find it an additional burden. Indeed, even in the fight against Nazism, 'the good war' in an American perspective, difficult moral choices had to be made. Historical analyses of the variety of responses that soldiers actually made clearly implies that there were choices within all military hierarchies even during a total war like World War II (Burleigh 2011, especially pp. 324–328, 360–393; Neitzel and Welzer 2012).

Modern military operations are justified more directly in moral terms. In the future it is likely to leave individual soldiers to judge their own participation as an individual moral choice. South African forces stressed the frustration they felt when forced to watch civilians harmed and villages destroyed without being able to do anything, and, even worse, when not allowed to do anything (Heinecken and Ferreira 2012, p. 47). Interestingly, unwise orders present harder tests for moral judgements than outright immoral orders. The reason is that there is a clear duty to disobey immoral orders, whereas more ambiguity surrounds unwise orders, and that poses a tougher test of the soldier's conscience (Sherman 2010, pp. 24–25). Particularly, *ambiguous* Rules of Engagement create 'confusion, vulnerability and disempowerment', as soldiers in modern peacekeeping are confronted with individual moral choices (Heinecken and Ferreira 2012, p. 48). This recent trend of moral justification may unreasonably burden the soldiers: 'it seems too much to expect soldiers on the verge of deploying, or having deployed, to track shifts in official rationales and to puzzle whether they are now, at this particular point, fighting for a cause that justifies the use of lethal force' (Sherman 2010, p. 45).

In today's international military operations, the intervening force nearly always enjoys the upper hand militarily. To find oneself in a situation where it is relatively easy to take life may be a source of stress for the individual soldier (Burleigh 2011, pp. 362–363). Moreover, soldiers report feeling tainted by association with a war that was flawed. The soldier did nothing wrong, and was at the extreme periphery of important decisions, but merely being associated with wrongdoing may be a burden (Sherman 2010, p. 52). They feel suckered, a stronger feeling than tainted. It is a feeling of finding oneself in the middle of something that one cannot believe in; in particular, it is the feeling of being misled or toyed with by those in charge, in a situation where one has intense feelings of loyalty towards the military forces and the state. Interestingly, the anger is often mainly within the military, directed up the chain of command, to seniors trusted not to squander the sacrifice the solider is willing to make (Sherman 2010, p. 54). That they did not speak out, *as was their duty as good soldiers*, gives rise to a sense of betrayal of the intense attachment that is a part of military life. Finally, disillusionment may occur if the soldier directly witnesses events that lead him or her to question the official rationale of the war. This leads to powerful emotions of triple betrayal. The soldier is lied to, in order for him or her to be a pawn in a dishonest domestic game, putting him or her unnecessarily at greater personal risk (Sherman 2010, p. 59).

In this book, high-intensity conflict is where the intervention force needs to escalate the use of heavy weaponry or weapons from other services, or where it takes part in firefights sufficiently often to dominate the operational everyday experience of the force. Moral factors shaping the intensity of the conflict are the perceived justness of the cause of the operation and the combat it leads to. As far as the responsibility for their own actions is concerned, mirroring the distinction between the mandate of the operations and its implementation, the shaping factors are the frequency and difficulty of morally challenging decisions encountered by the troops during the operation.

In modern international military operations, the forces may find themselves typically neither in low-intensity conflict nor in high-intensity conflict thus defined. They will find themselves in a conflict where they are constantly changing between a need for a peaceful *modus operandi* and a more belligerent one. This kind of constant change between two very different mind-sets and ways to operate puts great demands on each individual soldier and his or her unit, both mentally and in practical terms. It will be easier in a situation in which the troops mainly operate in a low-intensity environment but encounter isolated pockets of resistance or vice versa.

They will also find themselves in conflicts where their presence is justified differently than in conventional wars. This will be the case also for high-intensity operations where the actual fighting will resemble industrial war but the reason for being there is different. Modern international military operations put new operational demands on soldiers. Planning, rotation and follow-up need to take justification, individual experience of combat and change from combat to outreach into account.

Operational environment – from physical terrain to social terrain

The traditional definition of operational environment comprises all physical factors by which an intervention force is affected during its operation. An additional dimension that emerges from the new trends in previous parameters is that an international force increasingly needs to react to the local population. Therefore, a fundamental change in the operational environment parameter is the increased importance of knowledge of society, culture and attitudes, compared with importance of technical and material means (Kipp 2006, p. 12). In short, modern operations need to become intelligence-driven, and are moving in that direction (Norheim-Martinsen and Ravndal 2011).

In nearly all international military operations, the operational environment will be different from the conditions that the military forces are accustomed to train and operate in. For its opponents, however, it will be their home environment. The operational environment will therefore generally demand more adjustment for the international force than for its opponents, although the severity will vary from case to case. Environments favouring the international force do exist. The most important physical factors in an operational environment are geography, infrastructure, settlement pattern, topography and climate.

Geography matters in two contexts. The first is the distance from the states contributing resources to the area of operations. Large distances will demand strategic lift capability of forces, weapons supplies and other equipment. The second is the accessibility of the area of operations in terms of harbours and airports with sufficient capacity. For example, the access to Eastern Congo, a practically land-locked area with no direct access to the sea, is significantly more demanding than to a coastal area with nearby harbours and developed port facilities. Political conditions may reinforce geographical conditions. When the connection overland from Pakistan to Afghanistan is vulnerable to attack, or must pass through politically unstable countries in Central Asia, it strains resources and creates an uncertainty that may affect operations. Also, flying ammunition over third countries into the area of operations is regulated by permits and rules, for example, regulating the amount of ammunition per plane, making the operation more complicated.

The infrastructure in the area of operations may pose an even greater obstacle. International military operations typically take place in countries with weakly developed infrastructure. The problems are magnified if the military presence is needed in peripheral areas in which the infrastructure is even worse. Roads matter particularly. The construction and protection of roads, ports and airports may emerge as an important task in military operations.

Settlement patterns used to be considered of little importance in international military operations because the operations only rarely centred on urban areas. A dramatic increase in altered settlement patterns in large parts of the world may change this. A consequence may be that international military operations generally shift towards taking place in predominantly urban rather than predominantly rural areas, with the operational consequences that involves. Therefore, this book includes a separate chapter on urban operations.

Topography describes the dominating type of landscape: plains, savannahs or steppe, desert, forests or jungle, swamp or coastal area, mountains or cultivated land. Some kinds of landscapes favour irregular forces, others regular forces. Afghanistan's impassable mountain areas may hand the Taliban an advantage as they carry less equipment than do Western soldiers. Traditionally, jungles, and to some extent mountains, provide hideouts for insurgents, making it easier for them to temporarily withdraw from the fighting. Plains and deserts usually give conventional forces the advantage, particularly if they enjoy support from the air. The insurgents become visible and easily destroyed targets for long reach precision weapons. To some extent, to melt into the civilian populations is a countermeasure. In short, the problem for a Western force fighting an insurgency is usually *finding* the enemy, not prevailing when it does. Hence, a topography that is an obstacle to movement and provides places to hide, favours the insurgents.

The climate in the area of operations matters. Climate tends to favour the insurgents because, fighting on home turf, they are usually well adapted to their daily living conditions. Soldiers from other corners of the globe will more often than not face a new challenge of some kind. Previously, tropical diseases were a major problem, but today modern medicine can provide effective vaccines or prophylaxes against most of these diseases. Military equipment may be affected by the climatic conditions. Vehicles and weapons may have problems with sand, dust or the temperatures in desert areas, but most of this can be overcome by technical modifications prior to deployment or by additional maintenance during the operation, if the resources are made available. Even though there are remedies for climatic challenges, they divert resources from the operations and are an additional constraint on operational patterns. If the area of operations is at a high altitude, the reach and lift of helicopters are, for example, severely constrained.

The properties of the population have become an important part of the operational environment. These properties are usually referred to as *human terrain*, defined as '[t]he aggregate of socio-cultural traits present at a specific temporal, geo-spatial unit' (Eldridge and Neboshynsky 2008, p. 19). In modern operations, understanding culture and norms has become increasingly important. Behaviour offensive to local norms or policies that may be perceived as a threat to cultural values will result in local negative attitudes towards an international force and the regime it is trying to support. Knowledge of socio-economic conditions in a society is obviously crucial when economic and social means are used. The use of economic and social means can easily generate tensions that produce new conflicts instead of solving the old ones (Wilder and Gordon 2009). Knowledge of ethnic and other identities also matters because supporting one party in the conflict often leads to conflicts with the other.

To develop a strategy for 'non-kinetic' means in a conflict, it is necessary to realistically assess the reaction to various means in different segments of the population. Culture, norms, socio-economic and ethnic relations have therefore become an important part of the operational environment of an international force. Solid and relevant knowledge of these factors is crucial for the force to succeed in the mission. Knowledge of societal factors must be evaluated against environmental conditions unique to each individual operation. Last but not least, what is relevant information is determined by the objectives of the operation. Arguably, less ambitious goals in future operations could lead to less emphasis on knowing social factors. Therefore, to conclude, it is difficult to define the relevant societal parameters in a concrete way beforehand but that does not mean that their importance is diminished. What is known is that modern operations will continue to have objectives that affect the population, creating a need for information about the social make-up, fusing civil and military information (Norheim-Martinsen and Ravndal 2011).

Relative force composition and strength – from combined arms to moral anguish

In conventional high-intensity warfare, military force is the ability to compel the adversary to submit to our will. If your forces are able to compel, they are stronger than the adversary. Military forces can achieve strength by being numerically superior or by superior quality in some way. Weapons systems may support each other – 'combined arms'. There is more to relative military force than numbers. The adversary's attempts to protect themselves against one system may lead them to be more vulnerable against other systems. The natural defence for an armoured unit when attacked from the air is a combination of spreading the force out and digging in. If the air attack is combined with a massive assault with armour, the enemy is confronted with a dilemma. On the one hand, they may choose to spread the forces out, to minimise losses from the air, but then present an easy target for the adversary armour. On the other hand, they may concentrate forces to fight the incoming armour more effectively but become very vulnerable to attacks from the air. Thus, both relative force composition and numbers matter for military strength.

As discussed in the section on conflict intensity, the international force nearly always enjoys superior capability to escalate in this way. There is a trend in international military operations that the adversary not only has realised this but has found ways to mitigate Western military superiority in this sense: '...the enemy is deliberately choosing to keep the level and nature of the conflict where our advantages of numbers and equipment is neutralized' (Smith 2007, p. 278). The Western countermove, as argued by Arreguin-Toft (2006, pp. 38–45) is to make our forces more like the irregular method employed by our adversaries. There is evidence that this is the way for a Western force to prevail when faced with an opponent employing irregular tactics (Arreguin-Toft 2007). However, the objective in international military operations is not always to defeat an enemy in order to transform their society. Arguably, there is a move away from the large, society transforming objectives that at one point defined Afghanistan operations towards less ambitious objectives. The problem for Western forces is that it may not be enough to make one's forces more like the adversary's; large numbers are also required.

Military strength may be judged relative to the adversary or to the population in the area of operation. There is a trend towards using the latter. First, to go in with a sufficiently large international force is a basic precondition to succeed. In addition to the size of the population in the areas of operation, what constitutes sufficient strength also depends on the objectives of the operation. In traditional peacekeeping, where the parties have accepted the presence of the international military force, the task was surveillance of the parties. In these operations, there was often no need to use force at all. Historical experiences indicate that in peacekeeping operations four to ten soldiers per 1,000 inhabitants in the area of operations is sufficient. If the objective is to fight and prevail against an insurgency, troop levels need to be higher; about 20 soldiers per 1,000 inhabitants has often been suggested (Quinlivan 1995, 2003; Nardulli *et al.* 1999). The numbers are obviously approximate, particularly as the efficiency of local counter-insurgency forces may vary widely. The current trend for Western forces is to reduce their ambitions, settling for a less violent Iraq rather than the model democracy once envisioned. This trend seems likely to continue, leading to smaller Western forces in future operations. The avoidance of ground troops in OUP in Libya seems to point to a future with smaller numbers of troops.

Another point is that an international force is often unable to take losses without undermining domestic public support for the operations. This point explains why Western forces so often try to exploit their superior technological level and superior operational abilities, and emphasise force protection, ignoring the advice of Ivan Arreguin-Toft (2006, 2007) and David Galula (1964 [2006]).

Any counter-insurgency or stability operation in a populous country will be demanding in resources and manpower. To succeed, extensive local support will be necessary. To build and train such a force is therefore a vital task to achieve an acceptable force relative to the insurgency. There are two main ways to ameliorate the shortage of troops in these operations.

The first is to seek local allies. The advantage is that the number of troops available to fight the insurgency or to stabilise a given area increases while the area and population that needs to be protected often decreases. The downside is that these allies are local strongmen and warlords undermining the state and good governance. The second way is various forms of military assistance to build local forces. The advantage is that it is a vital part of the build-up of local government capacity, often an objective in these operations. The downside is that it is more demanding, especially in terms of time.

The next chapter considers how an international high-intensity mission may come about and analyses the future military implications of our first mission type.

References

Arreguin-Toft, Ivan (2006). *How the Weak Win Wars: A Theory of Asymmetric Conflict.* Cambridge: Cambridge University Press.

Arreguin-Toft, Ivan (2007). "How to Lose a War on Terror: A Comparative Analysis of Counterinsurgency Success and Failure". In Jan Angstrom and Isabelle Duyvesteyn, Eds. *Understanding Victory and Defeat in Contemporary War,* London: Routledge.

Bellamy, Alex J., Paul Williams and Stuart Griffin (2004). *Understanding Peacekeeping.* Cambridge, UK: Polity Press.

Boulden, Jane (2005). "Mandates Matter: An Exploration of Impartiality in United Nations Operations". *Global Governance* 11(2), 147–160.

Burleigh, Michael (2011). *Moral Combat: A History of World War II.* London: Harper Press.

Doyle, Michael W. and Nicholas Sambanis (2006). *Making War and Building Peace: United Nations Peace Operations.* Princeton, NJ: Princeton University Press.

Edelstein, David M. (2010). "Foreign Militaries, Sustainable Institutions". In Roland Paris and Timothy D. Sisk, Eds. *The Dilemmas of Statebuilding.* New York: Routledge.

Egnell, Robert (2011). "Lessons from Helmand, Afghanistan: What now for British Counterinsurgency?" *International Affairs* 87, 297–315.

Eldridge, Erik B. and Andrew J. Neboshynsky (2008). *Quantifying Human Terrain.* Monterey CA: Naval Postgraduate School.

Galula, David (1964) [2006]. *Counterinsurgency Warfare: Theory and Practice.* Westport, Conn: Praeger Security International.

Gentile, Gian P. (2008). "Misreading the Surge Threatens U.S. Army's Conventional Capabilities". *World Politics Review,* 2 March.

Heinecken, Lindy and Rialize Ferreira (2012). "Fighting for Peace". *African Security Review* 21(2), 20–35.

Huntington, Samuel P. (1957). *The Soldier and the State: the Theory and Politics of Civil–Military Relations.* Cambridge, Mass: Belknap.

Johnstone, Ian (2011a). "Managing Consent in Contemporary Peacekeeping Operations". *International Peacekeeping* 18(2), 168–182.

Johnstone, Ian, Ed. (2011b). *Peacekeeping's Transitional Moment.* New York: Center on International Cooperation.

Kipp, Jacob (2006). "The Human Terrain System: A CORDS for the 21st Century". *Military Review* (September–October).

Kuehl, Dale (2008). "Inside the Surge: 1–5 Cavalry in Ameriyah". *Small Wars Journal.* 29 October.

Marston, Daniel (2008). "British Operations in Helmand". *Small Wars Journal.* 13 September.

Marten, Kimberly (2006). "Warlordism in Comparative Perspective". *International Security* 31(3), 41–73.

Marten, Kimberly (2011). "Warlords". In Hew Strachan and Sibylle Scheipers, Eds. *The Changing Character of War.* Oxford: Oxford University Press.

Nardulli, B., K. Pollack, T. Szayna and B. Watts (1999). "Coup-proofing James T. Quinlivan". *International Security* 24(2), 131–165.

Neitzel, Sönke and Harald Welzer (2012). *Soldaten: On Fighting, Killing and Dying. The Secret World War II Transcripts of German POWs.* London: Simon Schuster.

New York University Center on International Cooperation, (2011). *Annual Review of Global Peace Operations, 2011.* Boulder, Colo: Lynne Rienner Publishers.

Norheim-Martinsen, Per Martin and Jacob Aasland Ravndal (2011). "Towards Intelligence-driven Peace Operations? The evolution of UN and EU Intelligence Structures". *International Peacekeeping* 18(4), 454–467.

Quinlivan, James T. (1995). "Force Requirements in Stability Operations". *Parameters* 25, 59–69.

Quinlivan, James T. (2003). "Burden of Victory: The Painful Arithmetic of Stability Operations". *Rand Review* 27(2).

Ricks, Thomas E. (2009). *The Gamble: General David Petraeus and the American Military Adventure in Iraq.* New York: Penguin Press.

Schadlow, Nadia and Richard A. Laquement (2009). "Winning Wars, Not Just Battles: Expanding the Military Profession to Incorporate Stability Operations". In Suzanne C. Nielsen and Don M. Snider, Eds. *American Civil–Military Relations: The Soldier and the State in a New Era.* Baltimore, MD: Johns Hopkins University Press.

Sherman, Nancy (2010). *The Untold War: Inside the Hearts, Minds, and Souls of Our Soldiers.* New York: W.W. Norton.

Simpson, Emile (2012). *War from the Ground up: The Twenty-first-century Combat as Politics.* London: Hurst.

Smith, Rupert (2007). *The Utility of Force: The Art of War in the Modern World.* New York: Knopf.

Thruelsen, Peter Dahl (2009). "The Comprehensive Approach in Helmand – From a Military Perspective". In Flemming Splidsboel Hansen, Ed. *The Comprehensive Approach: Challenges and Prospects.* Copenhagen: Royal Danish Defence College Publishing House.

UN Department of Peacekeeping Operations (2008). *United Nations Peacekeeping Operations: Principles and Guidelines* (Capstone Doctrine). New York: United Nations.

Waltz, Kenneth N. (1979). *Theory of international politics.* New York: McGraw-Hill.

Wilder, Andrew and Stuart Gordon (2009). "Money can't buy America love". *Foreign Policy,* 1 December.

3 Future high-intensity conflict out of area

A possible NATO counter-regime operation in Africa

Sverre Diesen

The number of international high-intensity military interventions and campaigns around the world after the end of the cold war has been limited. So limited, in fact, that when we also consider the fall of the Soviet Union and the disappearance of the threat of an all-out war in Europe, the rationale for maintaining forces for high-end warfare may seem slim, at least for most European countries. Indeed, General Sir Rupert Smith in his celebrated book *The Utility of Force* claims that this is definitely so and that war in the future will be amongst the people rather than between people – a concise way of saying that conventional inter-state war has made itself redundant, and that asymmetric warfare is the war of the future. The question is, therefore, whether the two Gulf Wars in 1991 and 2003 or the air campaign in Libya in 2011 are examples of a dying species of armed conflict or may reoccur more frequently than we think, given circumstances that could quite easily materialise. The presumption of this chapter is that we should be careful not to write off the possibility of such campaigns in the future but retain the political awareness as well as the military capability required to meet them, if and when they happen. In order to substantiate this assessment, we will consider a scenario in sub-Saharan Africa. The narrative will describe a crisis developing in an imaginary central African country where the international community, because of inept handling of the crisis in its initial stages, is drawn willy-nilly into what eventually becomes a full-blown campaign of high-end warfare on the African continent. The relevant points pertaining to each of the phases of the crisis will be discussed as the narrative unfolds.

Part I – Crisis development

Introduction – the political and economic situation in X

Outside intervention is usually the result of protracted failure of governance in weak states, which produces an untenable situation requiring some kind of international response. The path leading to such intervention is often fraught with considerations regarding the operational theatre, available resources and deliberations of various possible responses in the United Nations Security Council (UNSC). This scenario is based on a mixture of previous cases and is discussed within the context of contemporary and future trends in the parameters that define international operations.

The landlocked country of X in central sub-Saharan Africa is typical of the region both geographically and demographically. It is rich in natural resources, but the economy and infrastructure are not well developed. Poverty and corruption are endemic, and standards of health and education are low. Two years ago, a democratically elected but weak government was overthrown in a bloodless coup when a group of army officers from the dominating tribe seized power. The new regime is totalitarian and repressive, displaying many of the traditional symptoms of bad governance such as incompetence, corruption, cronyism etc. The economic policies of the ruling Revolutionary Council have so far made matters worse rather than better, creating high unemployment and hyperinflation.

After seizing power, the new regime has resorted to brutal oppression along ethnic lines, which has created a huge refugee problem for the neighbouring countries. The regime also pursues an aggressive foreign policy, particularly towards its eastern neighbour Y. The new regime's violation of accepted standards of governance has triggered economic sanctions as well as a weapons embargo, following a UNSC resolution. However, the effects of the economic sanctions are mainly felt by ordinary people, while the weapons embargo is rendered largely ineffective because of the porous borders between X and several other countries. Six months ago, reacting to repression and worsening living conditions, the minority tribe in the eastern border province rose in rebellion. The rebel movement demands secession from X and freedom either to become independent or to join the neighbouring country Y where this tribe forms the majority. The regime attempts to crush the rebellion with extremely harsh methods, thereby exacerbating the refugee problem and creating a catastrophic humanitarian situation in the eastern provinces. At the same time, the regime tries to foster national unity by threatening Y with war over a contested and oil-rich province on Y's side of the border, while at the same time accusing its government of aiding and supporting the rebellion.

The crisis escalates

The question of armed intervention in X to prevent further atrocities and abuse by the regime against its own population as well as outright aggression against Y is raised in the various assemblies of the international community (IC), such as the UN, EU, Organisation of African Union, etc. However, there is little consensus about what should or can be done, on grounds of practicality as well as principle.

At this stage, the chiefs of staff of a number of Western countries are tasked by their governments to provide assessments of the feasibility of effecting a regime change by the use of military force in X. Initial estimates made by the various defence staffs suggest that this would require the deployment to X of both airpower and ground troops to defeat the regime's forces. An air campaign alone would not suffice for both geographical and operational reasons. X as a theatre of operations for Western forces is fraught with difficulties because of adverse climatic, health, geographical and other conditions. In spite of this, even a limited Western expeditionary force would rapidly defeat the armed forces of X, due to superior technology, training, discipline, and overall quality. A prerequisite, however, is that the IC is able to enlist the support of another African country as a staging area for the campaign. The intelligence estimates are also unanimous in warning about the strong possibility of a prolonged COIN campaign against scattered fragments of the regime's forces as well as other stakeholders, following a successful overthrow of the regime and cessation of conventional operations.

Meanwhile, the debate continues in the UN and other international bodies, but little happens as the situation in X keeps deteriorating and tension between X and Y continues to grow. Troop movements close to the border on both sides are seen as sign of imminent conflict. Given the disparity between the two countries' militaries – the armed forces of Y are little more than a gendarmerie – there is nothing to prevent X from conquering the disputed province in the case of a war. At this stage, the prospects of unmitigated aggression and a war of conquest on the part of X going unopposed and unpunished put pressure on the opposing countries in the UNSC. On the question of armed intervention, they all move gradually from a *No* position to an *Abstain* or even *Yes*, although within limits and restricted to inserting a peacekeeping force in the border region. The major power sponsoring X is sufficiently worried about the situation to apply pressure on the regime to accept a UN peacekeeping mission. Consensus finally being achieved, a resolution to this effect is duly passed in the UNSC, authorising a peacekeeping force of 15,000 troops to be inserted in the border region between X and Y under a Chapter VII mandate with a double mission:

- preventing clashes in the border region between the armed forces of the two countries;
- protecting the civilian population in X 'under imminent threat of physical violence', specifying neither the minority tribe nor government forces.

It follows that the mission shall not in any way actively support rebel forces or take sides in the internal war. Both governments accept the presence of a peacekeeping force; however, rebel forces inside X declare that 'the armed struggle will continue'. This also provides the regime with the necessary excuse to pursue the repression of 'armed bands of terrorists inside the indivisible motherland' without interference from UN troops.

Over the next months, a peacekeeping force of 15,000 troops consisting of various ground force elements – infantry, reconnaissance, signals, engineers, logistic and medical support, as well as a helicopter detachment for transport and liaison duties, but with no heavy fire support – is deployed. The force is designated UNMIX and is primarily drawn from African, Asian and some Latin American countries, with only token European contributions. UNMIX comes on top of an already significant civilian UN presence in the country.

Erosion of the UN peacekeeping mission

Following the deployment of the UNMIX peacekeeping force, the situation in the border province of X continues to deteriorate. Although large-scale cross-border military operations of forces from X into Y is prevented, the UN force is too thin on the ground and lacks sufficient surveillance capability to prevent raids where refugee camps are attacked and innocent civilian refugees are killed. The regime in X claims that the camps are sanctuaries and training camps for terrorists, while the government in Y – with its significantly weaker armed forces – tends to turn a blind eye, secretly wishing the refugees would give up and go back across the border to X. At the same time, a vigorous campaign against 'terrorist strongholds' is conducted inside X, which includes massacres and extensive abuse of minority tribe civilians and other non-combatants. The leadership of the rebel movement is forced across the border to Y and sets up a provisional HQ in exile, recruiting new fighters among their tribesmen in the refugee camps. This is exploited by the government in X as proof of the refugee camps being 'terrorist sanctuaries'.

However, as UNMIX adapts to the conditions and improves its tactics and techniques, it becomes more effective in preventing both cross-border raids and outright atrocities by regime forces against civilians. This is considered a serious and unacceptable obstacle to continued operations in the region by the political and military leadership in X. Expelling the peacekeeping force, however, has a political price – essentially falling out with their chief great power sponsor – which, for the time being, the regime is not willing to pay. In an attempt to render UNMIX incapable of influencing the situation in the border region, therefore, the forces of X begin a campaign of systematic obstruction and harassment of the peacekeepers. In one such incident, a group of UN aid workers and their UNMIX escorts are taken hostage by militia forces sponsored by the regime, which denies responsibility for the outrage and blames it on the general lawlessness of the border region – 'the result of an insurrection for which the UN must also be held accountable'.

This leads to renewed calls in Western countries for armed intervention, the overthrow of the regime in X and the reinstatement of a legitimate government committed to democracy and the rule of law. This is opposed by those countries who point out that there is no political framework to build on in X, and that consequently the conflict must be resolved by applying pressure on the present regime. Thus, an international political stalemate over the situation in X continues. Encouraged by this, both the regular and militia forces become more aggressive and unrestrained in their harassment of UNMIX. The UN troops are increasingly forced on the defensive and driven back to their camps, limited to a self-protection posture as much by their lack of capability as by the terms of their mandate.

Three months into the operation, members of the irregular tribal militia commit the worst atrocity seen in Africa for years, murdering between one and two thousand civilians from the minority tribe in X. Units from the regular forces in X do not take part in the massacre but prevent a nearby UNMIX unit from intervening simply by threatening to engage it with heavy weapons if they try to move out of their camp. Barely two weeks later, a group of civilian UN aid personnel is taken hostage by an unidentified armed group, an action for which the government claims no responsibility. A number of nations contributing to the UN effort in X now signal that they will withdraw from the operation, as security for their deployed personnel can no longer be taken for granted.

Encouraged by the increasingly passive posture of the UNMIX forces, the government in X steps up the rhetoric directed against Y, and restates its claim to the disputed border province. Believing that there is minimal risk of international intervention, it also takes advantage of the situation and demands that the departing UNMIX contingents must leave all heavy equipment behind, such as vehicles, communications equipment, medical equipment, command and control infrastructure etc., otherwise 'the safety of the UN evacuees cannot be guaranteed'. The UNSC, amid universal condemnation of the actions of the government in X, meets to discuss the situation, with particular regard to securing the liberation of hostages and the safe evacuation of all UN personnel from X, military as well as civilian.

Confronted by the possibility of a complete UN loss of face and credibility, similar to that experienced in Rwanda, Bosnia and Sierra Leone in the 1990s, the UNSC now unanimously passes a resolution in which it commits itself to free the hostages, secure evacuation of all UN personnel and bring the perpetrators of crimes against humanity to justice. The government of X warns the UN that such an operation 'will be opposed by all available means by the armed forces of X, and will be defeated', but at the same time offers 'to prosecute certain irresponsible elements' in order to stave off intervention at the last moment. However, this is considered to be 'too little too late', and the deliberations in the UNSC about a new resolution containing the terms of reference for an armed intervention continue.

The UN is now faced with an obligation to intervene on behalf of its own personnel as well as to prevent the threatened genocide of the minority tribe in X and a war of aggression against Y. Despite the immense challenges this implies – including the likely resistance by the armed forces of X – the UNSC passes a final resolution. In this, the government of X is outlawed, and the UNSC calls upon the IC to bring about a regime change in X. However, many of the contributing nations to UNMIX declare that they are not prepared to support a counter-regime campaign against determined resistance from the comparatively strong government forces in X, since this is a totally new and different operation from the one they signed up to, not acceptable to their respective public opinions. It is also the assessment of Western intelligence services that such a task is well beyond the military capability of most of them, individually and collectively. Despite a sufficiently robust mandate, in other words, the limit to the power of the IC to enforce its will is set by the limited military capability of the original troop-contributing countries.

Realising this, feelers are put out by the UN to NATO, the African Union (AU) and the EU as possible 'contractors' for the required counter-regime operation. Despite their misgivings, the Western countries see that the future of the UN is at stake, and that their interests are not served by the world organisation being humiliated by a third rate African dictatorship. Despite comparatively few NATO members being involved in UNMIX, the North Atlantic Council therefore accepts the responsibility and issues an initiating directive to the Supreme Allied Commander Europe (SACEUR). At the same time, diplomatic feelers are put out to country W, situated on the Atlantic coast and bordering on X. The government of W, always apprehensive of the regime in X and the risk of ethnic conflict spreading, readily agrees and decides to put its military infrastructure at the disposal of the IC.

The government of X, faced by imminent intervention, is divided in terms of its strategy to deal with this situation. One group within the ruling council favours backing down in front of mounting international pressure and what must in the end spell defeat and indictment at the International Criminal Court (ICC). The hard core, however, which makes up the majority of the council, and to which belongs its 'strong man', claims that this is already too late, and that the only hope now is to take the entire UNMIX force hostage and hold the IC to ransom for their release, in return for immunity from prosecution.

Assessing the options for outside intervention

First of all, we should note that high-intensity war might actually emanate from the sort of internal strife or intra-state conflict which Rupert Smith calls 'war amongst the people'. The two kinds of armed conflict, in other words, are not mutually exclusive.

The end of the first phase of the crisis comes when an insufficiently equipped UN force ends up actually exacerbating the problem by becoming hostage to a ruthless regime, which may then hold the IC to ransom. Political bickering and unwillingness to face up to the many problems of a more robust intervention – while at the same time feeling that something has to be done – may result in a compromise leading to an even worse situation than the one which triggered intervention in the first place. Despite a more realistic conceptual approach to peacekeeping prevailing at the UN, as described in Chapter 7, with most operations receiving a Chapter VII mandate (Winther 2014, p. 12), the effectiveness of such operations is still in question, but now because the available forces lack the capabilities required to take advantage of a more robust mandate.

Consequently, in this scenario, with the credibility of the entire notion of some kind of world order on the line, political agreement is finally reached to apply sufficient military force to outlaw and oust the thoroughly discredited regime in X. This brings up the question of who should do the job, or more specifically: what is the role best played by the UN in such a scenario? Should they conduct the operation as well as legitimising it, or should they just provide legitimacy, and outsource the execution to some other organisation or agency? Quite often, governments highly supportive of the world organisation will advocate UN leadership, in order to bolster its image as the ultimate arbiter and authority on the world stage. However, there is much evidence to suggest that military operations over and above a certain level of intensity and complexity are better left to a dedicated security organisation or alliance, such as NATO. This is only natural, for a number of reasons. A permanent military organisation comes with a trained and workable command structure, with an agreed working language, standing operational procedures, established doctrine, forces that are interoperable and have trained together etc., – which adds up to a tremendous advantage compared to starting from scratch to establish all these things and fulfilling the mission at the same time.

These advantages are also likely to persist in the future. A UN force headquarters improvised from a mix of officers and NCOs from troop-contributing countries with no prior experience of working together will normally present significant procedural and cultural problems, the language barrier being only one of many examples. The UN should therefore see itself as the political overlord and taskmaster, setting the terms of reference and defining the goals to be achieved – and then let the professionals get on with the job. This will not detract from the prestige of the UN, as long as governments are reminded that professionalism in these matters will save both blood and treasure.

The next thing to consider, however, is the question of whether Western military forces are really the best possible means to serve the required political ends, even if they seem to be the only option in terms of general effectiveness. This is particularly so on most of the African continent, where conditions are not necessarily very suitable for operations with Western-type military forces. We need to remind ourselves that whenever Western forces have so far deployed out of area for high-intensity operations, it has been to the Middle East, which is significantly different from sub-Saharan Africa. The conditions in Iraq and other Middle East countries in many ways constitute 'a tactician's paradise', as was indeed the case in the North African campaign during World War II. The desert, with its open featureless landscape, its mainly hard surface allowing tanks and armoured vehicles to manoeuvre freely and its zero risk of collateral damage to civilians and civilian infrastructure, is tailor-made for the Western armed forces designed to oppose a Soviet invasion of Western Europe.

Sub-Saharan Africa, on the other hand, is anything but that. Add to this the lack of infrastructure to support operations logistically, the medical hazard and the rigours of the tropical climate, and we have an environment with which the armed forces of Western countries are not designed to cope. By the same token, such conditions are almost ideal for irregular or insurgency warfare, particularly by indigenous forces tuned to the conditions. Note also that the sort of cultural restriction on campaigning which we know from the Middle East in the shape of Ramadan has its climatic African parallel in the rainy season. This is why a UN peacekeeping or peace-enforcing mission with mainly African or possibly some Asian troop-contributing nations will normally present itself as the better solution in a crisis of this nature, given that their other shortcomings can be overcome or reduced.

Part II – Intervention and counter-regime operation by international forces

Preparations

Eventually, the UNSC passed a resolution authorising the use of force in X. However, the resolution did not specify the political end state in any detail, over and above bringing down the present regime, taking the indicted leaders to the ICC and establishing some kind of democratic government. However, NATO failed to solicit from the UN a description of how they envisaged the end state, identifying some sort of political party or organisation with sufficient legitimacy to act as an interim government. The Alliance had to work this out for itself, and eventually chose the People's Movement of X, a bi-tribal organisation in exile. This by and large legitimate and accepted group embraced the task, provided massive reconstruction support could be guaranteed following the end of hostilities. Based on this, NATO's planners could derive a consistent military end state as a basis for operational planning.

The outline plan of operations for the campaign in X named Operation COPPER DAWN was subsequently presented by SACEUR and contained four phases:

- In phase 1, deploy air power to bases made available in W, from which all air defence capabilities in X will be destroyed or suppressed, while at the same time deploying ground forces to assembly areas and forming-up positions close to the border with X.
- In phase 2, commencing on the completion of the deployment of ground forces, launch concentrated air strikes to destroy and disrupt the armed forces of X deployed to meet the invasion.
- In phase 3, launch a ground offensive across the border from W, breaking organised resistance by the armed forces of X, destroying all remaining offensive military capabilities and securing control of the larger capital area. An airborne task force was to take and secure the capital concurrent with the opening of the ground offensive, detaining the political leaders of X and securing important infrastructure in and around the capital.
- In phase 4, disarm and demobilise all remaining enemy forces, make safe weapons and munitions, secure military infrastructure and subsequently organise a cadre for a future indigenous security force based on captured materiel and personnel cleared of complicity in serious crimes against its own population.

The air contingent consists of a mix of multi-role fighters and strike aircraft, as well as un-manned aerial vehicles, tankers, reconnaissance, electronic warfare (EW) and airborne warning and control aircraft. The ground force totals three brigades, one mechanised, one light infantry and one airborne, with the relevant support units and under overall command of a divisional headquarters. For the first time in a NATO-led campaign, offensive cyber operations will be incorporated in the overall effort.

Deployment of the air and land contingents of a NATO force to W began as scheduled, which led to the immediate taking of all remaining UN personnel in X as hostages. These were subsequently held at various critical civilian and military installations as human shields by the regime, which at this stage had nothing to lose by such a step, triggering stern condemnation and warnings by the IC. Furthermore, the deployment of NATO forces to assembly areas near the W–X border led to massive redeployment of the armed forces of X, pulling back from the eastern provinces and the taking up of defensive positions along the two main routes leading from the border towards the capital.

The campaign

The air campaign of Operation COPPER DAWN was launched as planned on D-day, succeeding over the next few days in destroying all the MiG-23 fighters of the X air force, either in the air or on the ground, for no loss of NATO aircraft. Massive destruction of air defence assets, such as ground-based air defence systems, radars, communication nodes etc., was also effected, although a small number of SA-6 and SA-8 air defence systems were believed to have survived by remaining camouflaged and passive. However, on at least one occasion, a group of UN personnel held as hostages at an important radar site were killed or wounded in an air strike. On another occasion, the bodies of a group of UN soldiers were put on display after allegedly being killed in a NATO air raid on the national broadcaster in X, just as they were about to give a statement about 'the aggressive and unwarranted attack by NATO, endangering the lives of UN personnel'. The assessment of NATO intelligence, however, is that the soldiers were murdered by the regime for propaganda purposes.

Following the completion of the deployment to W and the air campaign giving NATO complete air supremacy, the ground offensive was launched after massive initial air strikes against the ground forces of X throughout their depth. The advance along the two main axes proved that manoeuvrability with tanks and armoured vehicles was severely restricted, whereas light infantry relying on basic small unit tactics fared much better in the close and heavily forested terrain off the main roads. At one stage, this necessitated the shift of main effort from the mechanised brigade's axis to that of the light infantry brigade. An airborne battalion, held in reserve, was also hurriedly committed to secure an important road and railway bridge between the border and the capital from being blown up by retreating X forces. The initial airborne operation against the capital succeeded in securing its most important objectives, including the international airport. It failed, however, to detain the key political and military leaders of the regime, who escaped and eventually reappeared in the eastern province declaring 'a ruthless guerrilla war against the colonial invaders'.

After a campaign lasting approximately eight days – subsequently known as the 200-hour war – conventional operations came to an end. The larger capital area was firmly under the control of NATO forces, although an insurgent presence in some parts of the city was already noticeable. Numerous enemy units, scattered by the rapid advance of NATO units, also posed an insurgency threat in the longer term. Although a number of UNMIX hostages were liberated by NATO forces capturing important objectives in and around the capital, several thousand still remained in the hands of the regime, held by the tribal militia under extremely harsh conditions in the eastern provinces. These were now beginning to be affected by diseases and malnutrition, as well as by abuse and maltreatment from their captors. However, a campaign in the underdeveloped eastern part of the country to liberate the hostages appeared to be an extremely hazardous undertaking, with far too few troops and lacking much of the essential equipment and training to operate under such conditions. In these areas, widespread harassment, abuse and outright persecution of the minority tribe at the hands of the irregular forces continued.

Of the three strategic objectives of the operation – preventing further massacres of the minority population and bringing the guilty parties to justice, preventing a war of aggression against Y, and liberating the UNMIX hostages – only one had been achieved. The regime still had a firm grip on the situation in the eastern provinces, the UNMIX hostages remained in their custody and the minority tribe was at the mercy of the militias as much as ever before. Only a war of aggression against Y seemed to have been prevented through the defeat of the regular forces of X. An extended campaign to liberate the hostages, protect the tribal minority and bring the members of the Revolutionary Council to justice seemed unattainable without massive reinforcements by forces trained and equipped to operate under the conditions prevailing in the eastern provinces during the rainy season. The end of conventional operations was duly declared, but without triumphant statements of 'mission accomplished', since this would have been patently untrue. In the UN, it became increasingly clear that a rapid defeat of the tribal militias and an extension of NATO's control to the whole country – including the detention of the regime's leading figures and liberation of the UNMIX hostages – were unrealistic goals, and the search for other ways of imposing the will of the IC started in earnest.

Aftermath and assessment

Most politicians, even without being familiar with Clausewitz' famous dictum that war is the continuation of policy by other means, will acknowledge that military force is an instrument to further political ends. However, a deeper understanding of the more subtle connection between means and ends, between the desired political end state and how force must be applied, is often lacking. This point is made abundantly clear by Rupert Smith in the introduction to his aforementioned book, *The Utility of Force*. During the age of Napoleonic warfare, the 200 years from 1789 to 1989 dominated by wars of annihilation – or the threat of such wars – this was scarcely a critical weakness. Wars largely suspended other political forms of interaction; they ended with the unconditional surrender of the losing side, after which the victor imposed more or less draconian peace terms at will – the payment of ruinous reparations, the ceding of provinces, the fall of dynasties (Huntington 1957). Today, with military force serving more limited political ends, there is a distinctive requirement for a better dialogue between politicians and the military about exactly what sort of political change force is required to bring about, and what kind of end state it is supposed to produce. Only then can the military planners sit down and work out a corresponding military end state, from which flows a consistent campaign plan. This calls for a very close dialogue across the political–military interface. At least some politicians at the national level need to understand strategy and the application of force, in the same way as other politicians understand the fundamentals of education, health or the economy. That senior military officers must understand how politics works at this level goes without saying (Cohen 2001, 2002; Schadlow and Laquement 2009).

Normally, the political end state will be couched in rather general terms, such as stability, peace, the rule of law, etc. This, however, is not sufficiently specific to be helpful when it comes to deriving a tangible military end state. Consequently, politicians need to be more explicit as to the details of the political settlement to be achieved. On the military side, there have historically been two types of objective: either the conquest of territory or the destruction of the enemy's armed forces. This has been in keeping with the tradition of the Napoleonic paradigm, where wars were about one or the other, if not both. Consequently, the military also needs to be specific when it comes to describing what it can do, in order to bridge the political–military communication gap. The restoration of recognised international borders, the eviction of the aggressor's forces from a given area within a certain time, the destruction of certain offensive military capabilities, the creation of an indigenous security force meeting certain requirements – all of these are examples of the specifics which need to be worked out between political and military leaders if force is to fulfil its role as a useful instrument of policy in an age of limited wars for limited political objectives. Not only are they decisive when it comes to designing the plan of campaign, they also form the basis for that much discussed but highly elusive concept – the exit strategy. However, even at the early stage of campaign planning, the prospect of a possible later counter-insurgency should be kept at the forefront of the planners' minds. We will return to this issue later.

Suffice it to say at this stage that a significant lesson, identified in both symmetric and asymmetric conflicts in which the West has been involved over the past 20 years, seems to be that we need a more realistic understanding of what we can hope to achieve in the type of armed conflict dominating in today's world. 'Victory' in its traditional sense – the condition in which the enemy can be forced to accept our terms unconditionally – seems to be a very elusive goal indeed. Instead, we should probably accept something less perfect as the best buy under the circumstances.

One aspect of this is the question of what to do with the villain of the piece – i.e. the contradiction between leaving a back door open for a reprehensible dictator and his henchmen to sneak out, and the ideal of the IC being seen to bring wrongdoers to justice. On the one hand, upholding certain standards of behaviour for governments and punishing evil-doers is obviously important, if the world is to progress in the direction we want it to. Indeed, the eventual prosecution of a criminal regime may be critical for the legitimacy of the whole intervention in the first place. On the other hand, as brought out by this scenario, this will also force a disreputable regime into a corner, with nothing to lose by burning its bridges and fighting it out to the last, with untold suffering and destruction as a result. There is no hard and fast rule to be derived from this dilemma, only an awareness of the problem and the need to strike a balance between the two, the right course of action depending on the circumstances and implications in each case.

So far, we have considered the higher political and moral aspects of military intervention. The importance of these notwithstanding, there are also the operational military practicalities to consider. Given the distance between North America or Europe and most countries in sub-Saharan Africa, military intervention in this part of the world is scarcely possible without another African country serving as a staging area and jump-off point for the campaign. In much the same way as the US has courted the regimes in Saudi Arabia, Qatar and Egypt in order to gain projection power to the Middle East and Central Asia, similar arrangements with friendly and reasonably politically stable countries in Central Africa may become indispensable. Preferably, these countries cannot be landlocked and must possess an infrastructure which will permit the landing of heavy strategic airlift, such as C-5 or C-17 planes. Its own armed forces must be capable of offering the necessary force protection to the expeditionary force, for reasons both of sovereignty and economy of force on the part of the coalition. Looking at the map and considering the various candidates, these are not trivial requirements, but Nigeria on the Atlantic coast, Kenya on the eastern seaboard and South Africa on the southern tip of the continent are probably the most likely possibilities, despite the obvious threats to long-term stability in all of them.

The nature of semi-regular warfare on the African continent, with its multitude of warlords, militias and warring factions operating with or against the regular armed forces, puts a premium on intelligence gathering and assessment. This applies equally to long-term strategic and short-term tactical intelligence. The former provides insight and cultural as well as political understanding of the theatre. The latter is crucial not only for ordinary operations, but also for such extremely intelligence intensive operations as hostage rescue, assessment of sympathies in various tribal or religious factions and the like. These are capabilities which the intelligence agencies and reconnaissance units of small and medium-sized European armed forces do not normally possess, and in relation to which agreements concerning the gathering and dissemination of intelligence between the partners of a coalition – never an easy thing to achieve – must be carefully worked out.

As the armed forces of most European countries have been shrinking proportionately to the technology-driven growth in the cost of military hardware, a number of capabilities necessary for power projection have come under threat. As demonstrated by the Libyan campaign – which was purely an air effort – such things as air-to-air refuelling, certain surveillance and EW platforms and strategic airlift are not available to NATO to the extent necessary for a serious deployment of forces and a subsequent campaign on another continent. The same thing applies to the necessary shipping to ferry tanks, artillery and huge quantities of supplies across the Atlantic, at least if the necessary tonnage has to be available at short notice. This is one of the critical issues when it comes to force planning in most NATO countries, since most of them are now getting close to the point where such capabilities are no longer affordable – at least not in sustainable numbers. The current situation, therefore, is that projecting a force even of just 3–4 brigades – a small fraction of the overall forces available to NATO on paper – is not possible without massive participation by the US. The solution to this problem seems to be increased and accelerated emphasis on various multinational force integration schemes, such as pooling and sharing and others. Such schemes are already in place in NATO for airborne warning and control and strategic lift (C-17), and must be considered for a whole range of other capabilities. The fact that current foresight studies in all major NATO countries see declining force structures as a pervasive trend extending well into the future underscores the importance of this.

As mentioned in the assessment of phase I of the crisis, Western-style armed forces are not designed to operate under the conditions prevailing in sub-Saharan Africa. Most of them are still organised, equipped and trained to operate in a mechanised role, whereas units designed for mountain, jungle, riverine or other environments are limited, often to the point where such expertise is only found in the small and exclusive special operations forces communities. However, the demand here is for a different heavy/light mix, with more emphasis on airmobile or amphibious troops and less on tanks and infantry fighting vehicles. Heavy mechanised units, although not necessarily completely unsuitable, face both tactical and technical limitations, as the conditions for sweeping armoured manoeuvres around the enemy's flanks and to his rear are rarely present.

Acquiring a capability to intervene militarily on the African continent, in other words, would exact large concessions from European force planners and designers, presenting them with some very difficult prioritisations at a time when financial constraints are already making their job difficult. In 1991, both France and the UK fielded divisional formations in the First Gulf War. That was possible, however, because they still retained the force structures of the cold war, which came to an end only months before Iraq's attack on Kuwait. Since then, practically all European countries have cashed in the peace dividend and completed a major restructuring and downsizing of their forces. Consequently, the same conclusion as the one pertaining to logistics applies equally to providing the actual forces: only the US will have the capability to mount operations on a scale over and above brigade level out of area.

We now turn to the design of the actual campaign plan. During the Kosovo conflict in 1999, it took NATO 79 days of intensive bombing from the air to force Serbia out of Kosovo, and halt ethnic cleansing and other atrocities. These crimes were mainly the work of Serb militias, as opposed to regular forces, such as the one led by the notorious Arkan and others. The Serb army units, with their tanks and other heavy equipment offering superb targets for NATO air strikes, could be dispersed and hidden under bridges, in barns or inside industrial facilities, with little regard for the possibility of a ground campaign, given the complete absence of NATO ground forces in the theatre. Had there been a credible threat of a land battle, however, the Serb army would have had to remain deployed and concentrated to meet that. This would have spelled their immediate destruction – not necessarily at the hands of NATO ground forces, but by the air strikes from which they would now have little protection. This example brings out the complementarity of the different arms and services of military forces, ancient or modern. An enemy confronted by a threat in a single dimension will have an infinitely greater chance of adjusting his own strategy – thereby rendering the threat ineffective – than one confronted by a multi-dimensional challenge.[1] This is why there is a need in most cases – including the intervention in X – for a joint and combined campaign. Although airpower is an extremely powerful and versatile element of any strategy, filling a number of different roles, there is no substitute at the end of the day for an air/ground – or possibly an air/sea/ground – combination of some sort. Otherwise, the scope for the enemy to respond to the threat by simply adjusting its own strategy and tactics is simply too great for the opposing force. This is all the more so in less industrialised countries in Africa, where infrastructure, production plants and other obvious targets for economically motivated air strikes simply do not exist.

NATO forces owe much of their efficiency to the fact that they operate on the basis of a largely common doctrine, taught in military academies and staff colleges, and serving as an intellectual framework for operational planning and execution throughout the alliance. The basis for this doctrine, however, is not the so far remote possibility of a high-intensity campaign in Africa, but a legacy of the cold war – a mechanised war in Europe against the field armies of the Soviet Union. This is, of course, as much of a handicap as the fact that their organisation and equipment – not surprisingly – are also reminiscent of that task, as previously pointed out. Preparations by NATO to enable a serious projection capability into Africa would therefore imply not just reorganising and reequipping a number of its forces to fit a different environment, but also developing new operational concepts laying the doctrinal foundation for such restructuring. As experienced by the British in the Falklands, this doctrine would probably have to reinvent classical small unit infantry tactics for fighting in forests and close country, albeit with the advantage of using more recent information, communication and precision-guided munitions technology.

The episode of the hostages being killed during the raid on the broadcasting facility in X brings out the challenge posed by the information age and the power of the media. As emphasised by most defence and security trend analyses, strategic information is no longer a supporting activity, but a line of operations in its own right. This is not in the sense that the media can be considered an active combatant, or expected to be supportive of any action taken by NATO forces – in fact, rather the opposite. The media will be highly critical, whatever their qualifications for assessing things fairly and objectively. Furthermore, they will not always understand the need for secrecy on grounds of operational security, unless explicitly briefed.[2] Hence the importance of 'the battle of the narrative' – the ability of the political and military leadership at both national and coalition levels to explain what they are trying to achieve and why, in a manner which is both understandable and credible to the general public, thereby reducing the space for rumours, speculation and deliberately falsified information. In this regard, and despite the enormous amount of attention and resources lavished on this subject in recent years, we still have a long way to go. This includes the need to clarify whether certain installations such as TV and radio stations serving the opposite side are in fact legitimate targets or – as the media themselves will claim – should have the same sort of protection as hospitals, religious sites or places of cultural significance.

Despite NATO supremacy across the spectrum of capabilities, war remains the province of Shakespeare's 'slings and arrows of outrageous fortune'. No plan will survive any amount of contact with the enemy, and chance remains a powerful influence on the battlefield. The campaign in X demonstrated clearly the need for competent staff work, beginning with the complex and time-consuming deployment itself. During operations, both Force and Task Force HQs had to adapt plans and issue orders as required by unfolding events. The switching of main effort from one axis of advance to another, with its many implications for fire support, logistics etc., is but one example; the decision to launch an airborne operation at short notice against an alternative drop zone is another. This is probably the most important reason why military campaigns and operations should be led by permanent, well-trained and integrated battle staffs, such as those provided by corps and similar HQs of NATO's force structure. To leave this to hastily improvised UN or other staffs is at best to promote inefficiency and at worst to court disaster.

One of the most frequently used political–military buzzwords or phrases of our time is 'exit strategy'. Rather than being something we start discussing when things begin to go seriously wrong, the criteria for when and how to get out of an operation should be part and parcel of the overall strategy from the very beginning. Since it is highly unlikely that Western forces will stay the course until countries we have tried to stabilise look like Switzerland, the key element of any such strategy in the future will be to establish some kind of indigenous security force. This would also be the case in the aftermath of the campaign in X.

However, one of the most important lessons identified in Afghanistan is that we should consider very carefully exactly what kind of security force is required – police or military. This depends very much on what kind of security situation we are faced with. A full-blown insurgency, with a well-organised and armed insurgent movement, will obviously require the establishment of regular armed forces, equipped and trained to respond in kind. However, a more nascent or potential threat may well be dealt with more effectively by a robust police force, capable of maintaining a presence throughout the country and prevent an insurgency from establishing an alternative authority in what might otherwise be a power vacuum. In the eastern and southern provinces of Afghanistan, where the Taliban has significant military capability, it has obviously been necessary to form an army – the Afghan National Army – with brigades and even army corps formations to handle the situation once NATO forces leave. In the Northern provinces, however, with their mainly Uzbek and Tadjik population, the Taliban as an essentially Pashtun movement has much less support, and hence no capability to mount military operations on a scale above what can be handled by a well-trained battalion. It is, however, a significant problem that the complete absence in many districts of a reliable and respected police force enables comparatively few Taliban organisers to walk into a village and requisition supplies, money or even recruits. The presence in a garrison 100 km away of a complete army brigade of three kandaks, or battalions, is of very little consequence in this situation, whereas a local police station with ten or twenty trained policemen would have been sufficient to see the Taliban off. Building an efficient security force is therefore not a question of having either an army or a police force. Quite often it will require a mix of a distributed police force maintaining law and order, protecting the population against extortion, abuse and parallel justice by the insurgents, coupled with small but highly mobile military units which can be committed against more serious and capable threats.

After the end of conventional operations, it was time for evaluation or stocktaking at the strategic level. This could not fail to deliver a sober realisation that two out of the three strategic objectives had not been met. The threat of a war of aggression by the old regime against the neighbouring country of Y had been averted, but a large number of UN troops and civilian aid workers remained hostages of what was now an insurgent movement in control of the eastern provinces and led by the same gang of rogue army officers as before. This also meant that the minority tribe in the same part of the country was still at the mercy of the tribal militias which had persecuted them throughout the rule of the previous regime. Furthermore, that this would be the situation at the end of a campaign to unseat the regime was entirely predictable, and to a large extent anticipated. The lesson, in other words, is that even the most robust and vigorous high-intensity response does not hold a promise of a quick and comprehensive political solution by force of arms alone. To fulfil the two remaining strategic objectives would require an extensive COIN operation in an environment very hostile to Western forces, at a price which public opinion in NATO countries would probably not be prepared to pay. So what went wrong?

The answer to that must be: not necessarily anything, as far as the campaign itself is concerned. The fact that a war and the subsequent destabilisation of the entire region have been avoided is a significant achievement in itself, going some way towards justifying the sacrifices and the unfulfilled intentions. In order to identify the mistakes, we must go back to the decisions made before it came to launching a full-scale military intervention. One serious mistake was probably to overlook some of the most important conclusions of the UN's own Brahimi report, and initiate an operation which was too weakly armed and resourced to achieve its mission, thereby setting it up for the hostage scenario which materialised (UN 2000). This not only put the lives of the UN personnel at risk, but also forced the hand of the IC by threatening to discredit the whole organisation, thereby putting paid to the only institution capable of restraining the forces of evil and imposing some semblance of order on an otherwise anarchistic world. Thus, an armed intervention became more or less unavoidable, despite its obvious prospective shortfalls.

The conclusion must therefore be that the military might of the Western powers, for all its sophistication, cannot offset the effect of political mistakes in dealing in time with rogue or failing states, at least not at a price which the populations of the liberal democracies of the Western world are willing to pay. Consequently, we must either solve the problems posed by these states using the economic and other non-violent means at our disposal, or we accept the fact that we may have to go all out with a military response to remove a regime which has become a threat to peace and stability. Furthermore, should we choose to do so, we must accept that this also entails a commitment to engage in the post-CORE phase, to support a new regime both militarily and economically for such time as it takes to make it self-supporting. This, at least, has been the approach of the Western world to such predicaments: a sense of responsibility for our actions embedded in the political culture, which implies a duty to put something better in its place, once we have taken it upon ourselves to remove an unacceptable regime. Or, to put it in the succinct, American parlance: 'If you break it, you own it.'

In neither case should we resort to half-baked military responses – UN-led or others – which will only exacerbate the problem. The instrumental tenet to be drawn from this is that a military response to a crisis cannot be scaled down all the way to zero, comfortable as that would no doubt be politically. Although force must be used in a way which is politically proportional to the challenge, it must be operationally capable beyond doubt of accomplishing the mission, leaving the other side in no doubt that they will lose if they choose to go to war. However, we should take great care not to end up in the opposite ditch, considering the words of Lt Commander Ron Hunter (Denzel Washington) in the film *Crimson Tide* from 1995: 'The purpose of war is to serve policy, but the nature of war is to serve itself'.[3] In other words, beware of the inherent dynamics of war, which has historically often meant that the military means take control of the political ends. The bottom line, however, must be that decisive use of military force still retains utility in today's or tomorrow's operations.

The following reconstruction phase, most probably with its organic counter-insurgency campaign, should be left as quickly as possible to the new indigenous security forces, the training and equipping of which must be part and parcel of any such strategy of intervention. This raises the question of whether purpose-made training teams to support the security sector in developing countries with legitimate and acceptable governments should be an integral part of the otherwise civilian aid packages offered to these countries – well before they fall victim to take-overs by the sort of people who took the country of X to destruction.

Notes

1 In the Libyan campaign, which at first glance might seem to be an exception to this rule, the ground element was provided by the insurgents, who brought Gaddafi's forces out into the open.
2 When the BBC speculated about an imminent attack on the settlement of Goose Green in the Falklands as such an attack was actually in the final stages of preparation, the commanding officer of the battalion assigned to the operation, Lt Col 'H' Jones VC of 2 Para, threatened in a rage to sue the BBC for manslaughter.
3 Also attributed to US political scientist and scholar Professor Richard Betts, of Columbia University.

References

Cohen, Eliot A. (2001). "The Unequal Dialogue: the Theory and Reality of Civil–Military Relations and the Use of Force". In Peter D. Feaver and Richard H. Kohn, Eds. *Soldiers and Civilians: The Civil–Military Gap and American National Security*. Cambridge, Mass: MIT Press.
Cohen, Eliot A. (2002). *Supreme Command. Soldiers, Statesmen and Leadership in Wartime*. New York: Simon and Schuster.
Huntington, Samuel P. (1957). *The Soldier and the State: the Theory and Politics of Civil–Military Relations*. Cambridge, Mass: Belknap.
Schadlow, Nadia and Richard A. Laquement. (2009). "Winning Wars, Not Just Battles: Expanding the Military Profession to Incorporate Stability Operations". In Suzanne C. Nielsen and Don M. Snider, Eds. *American Civil–Military Relations. The Soldier and the State in a New Era*. Baltimore, Md: Johns Hopkins University Press.
Winther, Dace (2014). *Regional Maintenance of Peace and Security under International Law*. New York: Routledge.
United Nations. (2000). *Comprehensive Review of the Whole Question of Peacekeeping Operations in All Their Aspects (The Brahimi Report)*. New York: United Nations.

Further Reading

Clausewitz, Carl Von (1832) [1984]. *On War*. Edited and translated by Michael Howard and Peter Paret. Princeton, NJ: Princeton University Press.

Diesen, Sverre. (2013). "Towards an Affordable European Defence and Security Policy? The Case for Extensive European Force Integration". In Janne H. Matlary and Magnus Petterson, Eds. (2013). *NATO's European Allies. Military Capability and Political Will*. Basingstoke, UK: Macmillan.

Janowitz, Morris. (1960). *The Professional Soldier: A Social and Political Portrait*. New York: Free Press of Glencoe.

Matlary, Janne H. and Magnus Petterson, Eds. (2013). *NATO's European Allies. Military Capability and Political Will*. Basingstoke, UK: Macmillan.

NATO. (2012). *Interoperability: Connecting NATO Forces*. www.nato.int/cps/en/nato-live/topics_84112.htm?

Smith, Rupert. (2005). *The Utility of Force: The Art of War in the Modern World*. London, Penguin Books.

Østerud, Øyvind and Asle Toje. (2013). "Strategy and Risk Assessment in NATO". In Janne H. Matlary and Magnus Petterson, Eds. (2013). *NATO's European Allies. Military Capability and Political Will*. Basingstoke, UK: Macmillan.

4 Counter-insurgency operations revisited

Robert Egnell and David H. Ucko

One of the more striking characteristics since the end of the cold war has been the plummeting concern over the prospect of great power conflict and the concomitant rise in various types of international intervention by these same powers in other countries' civil wars, counter-insurgency campaigns, and complex humanitarian emergencies. The liberal international context within which these operations took place has also increased the complexity of their political aims, as ambitions have expanded to include not only the creation of stability, but also of democracy, economic development and respect for human rights.

Conventional armed forces have struggled to adapt to this new context, while the political and military leaderships have also faced great difficulties in translating military action into political aims. This process has been particularly fraught when interventions have been launched in the midst of an ongoing conflict or where security is violently contested. The culmination of Western military experimentation with such interventions since the end of the cold war was the re-discovery of counter-insurgency as an operational approach to irregular armed violence and achieving a sustainable and legitimate political end state. Popularised as a means of solving the messy aftermath of otherwise successful conventional invasions, first of Afghanistan and then of Iraq, the focus on counter-insurgency has since led to a proliferation of both doctrine and thinking – in no small part based on the historical experiences of the British and French armies in the colonies. However, just as quickly as the concept rose from historical obscurity to be celebrated as having pulled Iraq back from the brink, it then fell from grace due to its record in Afghanistan. Intervention in foreign conflicts is now increasingly framed as anathema and counter-insurgency, if it were ever accepted as a helpful theory, is on its way out.

Based on the rise and decline of counter-insurgency over the past few years, this chapter assesses the future utility of this concept as a defence priority and area of research. The chapter lays out the limitations of the counter-insurgency approach while at the same time highlighting the importance of salvaging the hard won lessons of the last decade. More specifically, it does so with reference to the parameters this book establishes for the contemporary operational environment – the operational mandates, the issue of local consent, the relative force composition of the belligerents, and the likely conflict intensity.

In tackling this issue, the chapter must also grapple with the overriding or most dominant lesson from past counter-insurgency campaigns, namely that these ambitious experiments in state-building must be avoided. In the face of sub-optimal results in both Afghanistan and Iraq, Western interventionist powers have looked for and implemented supposedly 'cheaper' forms of engagement that protect their forces from the challenges seen in recent counter-insurgency campaigns. Even so, this chapter argues that several factors point to the continued salience of counter-insurgency. Indeed, the global trend of urbanisation, the West's enduring superiority in conventional combat, the attractiveness and effectiveness of irregular tactics to militarily inferior adversaries, the increased frequency of international state-building and the securitisation of 'state failure', particularly following 9/11, all point to a future of operations being conducted among civilians and, most often, with the objective of building government capacity. So while Western armed forces may not engage in operations on the scale of Afghanistan or Iraq in the near future, campaigns to come will in all likelihood call for similar skill sets and capabilities – not least at the tactical level.

The chapter starts out by defining the concept of counter-insurgency and then discusses the principles and requirements of counter-insurgency operations. Thereafter, the possibilities and constraints of future counter-insurgency operations are discussed. Given that large-scale counter-insurgency operations are unlikely to be politically viable in the near future, the chapter also highlights three options for international operations beyond counter-insurgency. What this section shows is that while counter-insurgency may be avoidable for third-party interveners, the complexities of warfare are here to stay.

Counter-insurgency – what's in a concept?

Any study of counter-insurgency must establish the criteria by which separate campaigns across time and space are brought together under this particular rubric. The question is germane, as counter-insurgency is often perceived to be a fluid concept that is entirely dependent on context (Kilcullen 2009, p. 183). From an analytical standpoint, this fluidity is entirely understandable but makes it difficult to pin down the exact nature of counter-insurgency and to distinguish it from other types of approaches or campaigns. Could we perhaps just as well call it stability operations, peace operations or asymmetric war? In the absence of such delineations, is counter-insurgency even a valid or meaningful term? How do we *know* that the campaigns in Iraq, or Afghanistan, were counter-insurgency campaigns and how will we recognise future such campaigns if and when they appear?

Faced with this semantic hurdle, the first recourse is to definitions. The British Army (2009, 1–6) defines counter-insurgency as 'those military, law enforcement, political, economic, psychological and civic actions taken to defeat insurgency, while addressing the root causes'. Insurgency, meanwhile, is defined as 'organised, violent subversion used to effect or prevent political control, as a challenge to established authority' (British Army 2009, pp. 1–5). When these definitions are taken to their logical conclusion, counter-insurgency emerges as any activity that purports to counter organised, violent attempts to obstruct the established authorities from asserting their control, and which also addresses the root causes behind the conflict. This is a helpful start, but difficulties immediately surface. First, who were the 'established authorities' in Baghdad and Basra in 2003 and 2006, or in Kabul or rural Helmand in 2001, 2006 or 2012? Second, what do we call counter-insurgencies where root causes are not adequately addressed, either as a matter of policy or as a failure thereof? Who identifies root causes and, even where properly determined, are they even central to the cycle of violence once it has fully taken hold (for an elaboration on this question, see Woodward 2007)? Furthermore, because both definitions are so inclusive, they still do not help us determine where counter-insurgency begins, where it ends and how it relates to other similar activities known by different names.

Ultimately, given the *sui generis* nature of most campaigns, the contested nature of most terms and the ambiguity and overlap of their respective definitions, it seems unlikely that greater intellectual investment in semantics will achieve the sought-after exactitude. To gain a slightly more specific understanding of the problem and help settle the criteria for inclusion, it may help to *describe* these types of campaigns rather than define them, to study them based on their shared characteristics rather than by what they were called. On this basis, the operations of concern to this study share three characteristics:

- the deployment of foreign and/or local security forces among a civilian population;
- the use of military force as a subset of a broader programme of military and non-military reform, aimed at addressing the causative factors of violence;
- an insecure operating environment, in which counter-insurgency forces, their allies or the population are regularly targeted.

The combination of these three characteristics unites the historical counter-insurgency campaigns of the colonial powers, the more recent experiences in Iraq and Afghanistan and, it is likely, most future land operations. Interestingly, these three attributes also tend to feature in what are termed 'stability operations', 'robust peacekeeping' and 'counter-guerrilla wars' and there is certainly no reason to exclude these types of engagements merely because they have not, often by mere happenstance, earned the label 'counter-insurgency'. All approximate to 'armed reform' (Marks 2009b, p. 20), an apt description of counter-insurgency. By implication, it should also be clear that the relevance of counter-insurgency experiences extends far beyond this particular term, touching upon what the authors perceive as rather typical, rather than atypical, complexities of political violence.

The principles and requirements of counter-insurgency approaches to operations

Having identified the types of operations of concern to us, the next question is what makes counter-insurgency so important in relation to the contemporary and future challenges of international operations. This is a pertinent question, particularly given the dissent surrounding the term, its practice and doctrine (Cohen 2010; Dunlap 2006; Gentile 2011; Luttwak 2007). Simply put, the modest yet at the same time crucial value of counter-insurgency lies in the challenge it poses to the many preconceptions about war that have tended to dominate Western strategic thinking. Specifically, in its principles, enumerated in most field manuals and key texts on the topic, counter-insurgency provides a corrective to the view of war as militarily decisive, apolitical and wholly distinct from peace. While this reductive view of war is most closely associated with US strategic thinking, the widespread Western use of terms like 'conventional' and 'traditional' (to describe state-on-state war) and 'unconventional' or 'irregular' (to describe insurgency and terrorism) reveals a more pervasive bias as to how wars are expected or preferred to unfold (Buley 2007; Cassidy 2005; Echevarria 2004; Kagan 2006; Weigley 1973).

How do counter-insurgency principles improve on this normative understanding of war? In essence, they hint at the challenging and at times counter-intuitive nature of the modern, populated and intensely political battlefield (Smith 2005). While there are now several lists of counter-insurgency principles in circulation, a quick examination of these five main principles helps make the point: political primacy, close civil–military cooperation, intelligence, minimum force, and flexibility (Alderson 2010; Mockaitis 1995; Rigden 2008). These five are the principles that have remained consistent throughout the centuries and that underpin doctrines, field manuals and theoretical texts on the topic (Cohen *et al.* 2006; Kitson 1977; Thompson 1966).

To the casual observer, these counter-insurgency principles will appear largely self-evident – and not just within the specific context of counter-insurgency. For instance, anyone well versed, or even superficially familiar, with the writing of Clausewitz will appreciate that war is at its heart political and should be informed and limited by the political interests for which it is fought. Similarly there is nothing particularly controversial about the need for good intelligence, and it is also clear that, where adversary and civilian look alike and intermingle, obtaining good intelligence will require a special understanding of and with the local population. It is equally difficult to find fault with the notion that a greater understanding of the environment, its people and structures will present external actors with more and better options, or that controlling and influencing key populations will first require that they are adequately isolated from the intimidation, threats and coercion of others. Operations involving a broad range of actors also require coordination and cooperation, and civil–military cooperation has therefore become a key principle in all operations since the end of the cold war. As to the focus on legitimacy, this is a fairly obvious corollary of the need to establish political and military control over select populations, whose consent greatly facilitates this task.

Still, though many of the principles would seem banal, they nonetheless appear necessary in illustrating the unique logic of counter-insurgency and its distinctiveness from the 'conventional' types of war for which most Western militaries train and prepare. For military institutions such as these, beholden to the utility of force as a stand-alone solution to security threats, the counter-insurgency principles represent a powerful antithesis.

Careful consideration of the principles and the challenge of their application helps prepare soldiers for the likely difficulties encountered in theatre. Gen. (ret.) Sir John Kiszely, former director of the British Defence Academy, provides a compelling summation: troops must possess the ability to

> apply soft power as well as hard...; work in partnership with multinational, multiagency organisations, civilian as well as military...; master information operations and engage successfully with the media; conduct persuasive dialogue with local leaders...; mentally outmaneuver a wily and ruthless enemy; and, perhaps most often overlooked, measure progress appropriately.
>
> (Kiszely 2007, p. 8)

As Kiszely adds, these competencies require an understanding of

> the political context; the legal, moral and ethical complexities; culture and religion; how societies work; what constitutes good governance; the relationship between one's own armed forces and society; the notion of human security; the concept of legitimacy; the limitations on the utility of force; the psychology of one's opponents and the rest of the population.
>
> (2007, p. 8)

This list of challenges and requirements goes some way toward explaining the chequered track record of historical and contemporary counter-insurgency campaigns – endeavours that are typically costly in both blood and treasure. Indeed, it is important to note that counter-insurgency will always be notoriously difficult, no matter how good the theory might be. It is relatively easy to derive priority tasks from past operations, yet knowing how to sequence, prioritise between and implement these represents a far more challenging proposition. It should not surprise, therefore, that while past campaigns reveal the *general* validity of a number of broad principles, the campaigns themselves were not always so successful.

The challenge of applying counterinsurgency approaches in the twenty-first century

Theories and concepts should be used to make sense of a complex reality and to support the dynamic strategic process of analysis, decision-making and implementation. This is not just an intellectual exercise: the concepts we use have an impact on how we interpret the conflict, prioritise resources and conduct operations. Selecting a concept, or a term (like counter-insurgency), requires great care: ideally, it should help us understand the true nature of the problem and how best to deal with it.

How does counter-insurgency measure up to this utilitarian approach within the contemporary and future strategic contexts? The concept has clearly been very useful in moving many armed forces from their exclusive focus on conventional warfare, yet, in itself, the idea of counter-insurgency has served better as an antithesis to past pathologies than as a prescriptive guide for ongoing campaigns.

The first challenge stems from the misinterpretation and overgeneralisation of the lessons of past counter-insurgency campaigns. Historians and military thinkers often stress the limited generalisability of operational and tactical approaches from one context to the next. One would therefore assume that when a colonial policing approach was revived to support the state-building campaigns in Iraq and Afghanistan, great care would be taken to appraise the differences separating these two worlds. Yet such analysis was all too rare.

Instead, the contemporary understanding of counter-insurgency was left at the mercy of a rather problematic reading of history. An increasingly recognised misstep in the contemporary reading of past theorists is its exaggeration of the 'hearts and minds' aspects of operations and its neglect of often equally and at times more important coercive components (Dixon 2009; Gumz 2009). Too often, polished historical accounts and memoirs from past campaigns were left critically unexamined, and instead a liberal twenty-first-century filter was applied that simply reinforced pre-existing biases. In fact, collective punishment, executions and forced population movement are but a few examples of past tactics, employed even in the most revered yet academically abused campaign – Malaya. That much of this scholarship was benignly intended to reverse the military's prior over-reliance on military force as a strategic solution helps explain what happened, but does not change its unfortunate outcome. The pendulum has now swung from one extreme to the other, and will continue to do so unless greater historical rigour is applied (Ucko and Egnell 2013).

There are also a number of important contextual differences between past counter-insurgencies and contemporary campaigns that should inform any attempt to transplant past approaches to the contemporary context. Past counter-insurgency operations took place as 'internal' challenges within the realms of empire; today, operations are typically conducted by coalitions and in support of weak yet legally sovereign and fully independent states (French 2011; Mackinlay 1998, p. 25). Despite the obvious scope for divergence, contemporary counter-insurgency doctrine still presumes sufficient harmony of interest between intervening and host-nation governments or at least an ability to push the latter toward the 'correct' course of action. Actual practice provides a more sobering perspective: in Iraq, the institutions either collapsed through war or were dismantled through coalition decree, leading to the infiltration of sectarian elements into positions of central power and a government whose interests often ran counter to those of the intervening coalition. In Afghanistan, the counter-insurgency campaign confronted a deeply dysfunctional state bureaucracy and a NATO headquarters that lacked the capacity and resources to run anything but the security aspects of operations. In both campaigns, difficulties with the host nation government were compounded by differences among coalition partners regarding approach, commitment, and contributions.

A further change has already been hinted at: the availability and competence of civilian means. The strategic intent in Iraq and Afghanistan would have required a substantial civilian component, large and capable enough to compensate for whatever weaknesses were found in-state. These resources were at the disposal of past empires, in the form of colonial administrations with local experience and understanding and local police forces that could provide public order (Kitson 1971; Rigden 2008, p. 13). Today, the political and civilian components of counter-insurgency are tremendously under-developed, despite efforts like the Stabilization Unit in the UK and the ill-fated Office of the Coordinator for Reconstruction and Stabilization within the United States. The result of this deficiency is a distinct mismatch between ambitions and resources.

The attempt to transplant past counter-insurgency approaches onto contemporary state-building efforts also risks neglecting the essentially conservative nature of counter-insurgency. The concept of 'counter-insurgency' presumes that the problem at hand is an insurgency that challenges the status quo. While successful counter-insurgency campaigns have often involved certain political concessions, counter-insurgency operations are fundamentally predicated on the survival of the state or its peaceful liberalisation by pre-empting more violent agents of change. The question is to what extent future international operations will allow for the conservative approach that counter-insurgency involves.

Iraq and Afghanistan highlighted the consequences of the clash between a strategy of regime change and state-building, and counter-insurgency as an operational approach. Given the fact that external coalitions toppled the existing regimes and instigated revolutionary societal changes in both countries, it takes a stretch of imagination to argue that the status quo was being protected or even gradually reformed. Instead, in these cases, the international community is better viewed as the revolutionary agent of change, and branding these efforts as counter-insurgency misframed the actual roles and contributions of different actors within those respective societies, not least those of the intervening forces. Perhaps most critically, it has led to an all too militaristic and optimistic view of what it would take to transform these societies with limited means and within limited time frames (Egnell 2013).

To discuss the extent to which counter-insurgency is likely to be a useful approach in future operations, we will now turn our attention to the parameters established for this volume and test them against counter-insurgency approaches to operations.

Counter-insurgency and the parameters for the future operational environment

The changes in the parameters for future operations established in Chapter 2 of this volume further highlight the tensions, while at the same time pointing towards the utility of counter-insurgency approaches in future international operations. First, the changes in mandates from stability through monitoring and observation to contemporary operations involving protection of civilians and societal change may at first glance resonate substantially with counter-insurgency approaches. Addressing political grievances and protecting the local population are indeed at the very heart of counter-insurgency activities designed to establish and maintain popular support. However, there is a key distinction to be made when these tasks are approached by a third-party actor operating on an expeditionary basis in foreign states and polities. In such contexts, greater attention must be paid to the strategic purpose of the undertaking: is it to resist a violent overthrow, to ensure continuity with the past or to mediate peaceful change? Each suggests the need for a radically different force posture, strategy and civil–military balance. If reform needs, as it so often does, to co-opt sources of resistance, how can a third-party push for such change without alienating the host-nation government whose consent for cooperation is necessary for progress to be made. These are questions with important implications for the military and its role in the field but – even more so – for those political and diplomatic agencies with key roles to play in future campaigns.

The discussion on the second parameter of consent highlights a general trend from legality to recipient legitimacy. However one may feel about the purported linearity of this shift, building and maintaining consent – of the host-nation government and its people – are key counterinsurgency considerations. General Stanley McChrystal's tactical directive to all units in Afghanistan as he assumed command is telling:

> Our strategic goal is to defeat the insurgency threatening the stability of Afghanistan. Like any insurgency, there is a struggle for the support and will of the population. Gaining and maintaining that support must be our overriding operational imperative – and the ultimate objective of every action we take.
>
> We must fight the insurgents, and will use the tools at our disposal to both defeat the enemy and protect our forces. But we will not win based on the number of Taliban we kill, but instead on our ability to separate insurgents from the center of gravity – the people. That means we must respect and protect the population from coercion and violence – and operate in a manner which will win their support.

However, the parameter also discusses the issue of host-nation consent, and, as this chapter has highlighted, conducting expeditionary counter-insurgency operations in support of a host government entails numerous challenges. First, winning legitimacy and support from the local population is inherently difficult for a foreign force, not just because of the readily exploitable symbolism of occupation, but also because a foreign force must win legitimacy for *another* actor: the host-nation government. Second, this level of commitment also means that friction between the external counter-insurgents and the host government is inevitable – not only in terms of the vision for the process, but also because of the limited sense of sovereignty that inevitably arises from the long-term deployment of foreign troops. The continued deterioration between President Karzai and the US administration is an obvious case in point. In sum, the population-centric counter-insurgency approach does not solve the challenges presented by the need for strategic and tactical level consent. Indeed, in the case of external intervention, the duration and scale of counter-insurgency operations may well have the opposite effect.

The third parameter is that of conflict intensity, which not only involves traditional intensity measured in numbers of troops, battles or bullets fired, but also the important variable of psychological and moral anguish. Counter-insurgency operations cannot be neatly placed along the continuum of conflict intensity, not only because of the diverse tactical activities involved, but also because it is in its character so dependent on enemy activity and strength. While counter-insurgency operations are certainly far from industrial interstate wars, at the tactical level troops are still likely to experience immense combat intensity from time to time. One needs only to consider the strength of Fuerzas Armadas Revolucionarias de Colombia (FARC) in Colombia, or of the Liberation Tigers of Tamil Eelam in Sri Lanka, to gain an appreciation of how widely 'counter-insurgency' practice can vary, even within individual conflicts. The nature of the insurgency tactics may also mean increased psychological anguish, as the source and timing of threat and danger are often unknown or unclear. Many soldiers in Afghanistan and Iraq experience intense 'battle' and the death of unit members without ever seeing the enemy. The fact that expeditionary counter-insurgency operations will always be wars of choice, that troops will face difficult moral choices while operating in the midst of the local population, and that these troops are often, regardless of their acts, seen as 'occupiers' rather than liberators in the eyes of the local population has the potential to reinforce the psychological anguish.

The parameter of the operational environment involves a discussion of physical and social terrain as well as of the instruments employed to control these different landscapes. Countering insurgency movements has traditionally emphasised the importance of population control. This, however, does not mean that the social terrain has replaced the physical terrain as the main battle space. Instead, in order to control the population, the physical and social terrains are both essential and mutually constitutive. When speaking of the social terrain, the issue of influence operations and 'non-kinetic' means invariably arises. How can the local population be influenced to support the host government and the international presence, and what is the military role in that process? This chapter has already highlighted the misplaced emphasis on winning 'hearts and minds' as a purely humanitarian endeavour. Another unhelpful tendency is for military organisations to distinguish between what they see as kinetic operations with the purpose of defeating the enemy and hearts and minds activities with the purpose of gaining local support and legitimacy. Rather than seeing all aspects of operations as influencing the local population, this approach created an artificial bifurcation – something that has proved problematic. Military organisations are experts in the use or threat of force to accomplish their objectives, which should therefore also be seen as their primary tool to influence the local population, or to 'win hearts and minds'. The question is not whether or how much force is used, but how it is used and to what political and social effect.

Regarding the parameter of relative force composition and strength, the very premise of expeditionary counter-insurgency operations is one of military superiority against a weaker enemy that seeks to off-set its weakness by engaging instead on its terms (Smith 2005, p. 278). The population becomes a critical player: as camouflage for the insurgents, as participants with agency and coping mechanisms, and as victims of both insurgent intimidation and state overreach. The general counter-insurgency wisdom states that about 20 soldiers per 1,000 inhabitants are necessary to control the population (Nardulli *et al.* 1999; Quinlivan 1995, 2003). What this statistic, tentative at best, does not acknowledge is the critical role of the population itself in providing for its own security. Indeed, there are precious few cases where counter-insurgents have successfully 'provided security' – the force ratios necessary for such benevolence call either for a fully developed police force (rarely found in conflict environments) or for a force deployment so large as to be unsustainable. The implication for expeditionary counter-insurgency actors is that they must not only ensure sufficient capacity – one of the chief lessons of the post-invasion phase in Iraq – but also learn how to identify and exploit opportunities for local partnerships, raise and develop local security forces and find common objectives around which cooperation is possible (draws on Ucko 2013). Such work calls for analytical skills, local understanding and strategic acumen – areas where, to date, success has been too contingent on improvisation, luck and circumstance.

Counter-insurgency is dead; long live counter-insurgency

As a result of these and many other factors, discussions of the 'future of counterinsurgency' must today acknowledge that counter-insurgency does not seem to have a future. Due to the changes in operational parameters, operational upsets in Afghanistan and Iraq and the semantic divisiveness and vagueness of the term itself, counter-insurgency – if it was ever was an appropriate lens through which to understand the security challenges presented by failing states – is no longer de rigueur. It has proved too costly – politically, financially and also in blood. This has engendered a push within Western capitals towards more affordable, risk-averse and, even, effective 'solutions' (as discussed in Chapter 1). Yet while expeditionary powers may have the luxury of avoiding counter-insurgency (in contrast to the states were insurgency is a daily, unavoidable and existential threat), the complexity of war remains and is difficult to change – so long as global ambitions are to be retained.

There is therefore something wholly unhelpful about the dogmas that surround the conversation about future counter-insurgency, with some suggesting it is 'dead' and others lamenting future engagement lest it becomes 'another Iraq' or 'another Afghanistan'. For sure, no one in their right mind would want to repeat these past campaigns, but then neither the Iraq nor the Afghan war began as counter-insurgencies. Instead, it was precisely our refusal to anticipate and prepare for the complexity of war and the enemy's ability to adapt that produced the problems faced in the last decade. There is therefore a certain hubris to the Western insistence that it will now place preconditions on the operational environment, or simply abandon it when it no longer suits. Nothing here condemns us to endless encores of the Iraq and Afghanistan campaigns, but nor can we return to the military thinking about war that dominated before these campaigns: a vision of war as an apolitical, militarily decisive and techno-logically driven phenomenon, unfolding on an isolated battlefield.

To do better in the future, we must think more creatively about how to engage with war's complexity and political essence in order to shape global security affairs yet *without* repeating the traumas of the last decade. Recent history suggests three ways in which this balance, in the future, is to be struck: the Libya model, the indirect approach and contingency operations in support of regional and international organisations. These three models have already been adopted in the last few years, and while they have been helpful in some contexts, they have also shown themselves to be highly reliant on key conditions and capabilities. Most critically, each requires far greater clarity about the nature and demands of expedition-ary operations, their typical duration and the challenges of operating as one member of a larger team. In other words, many of the hard-won lessons from the counter-insurgency operations of the past decade remain as relevant as ever, even if the *approach* has changed. It should be noted that the indirect approach will also be covered in later chapters on special operations, and advising and assisting.

The Libya model – death from above

Following weeks of civil war in Libya in 2011, pitting a rag-tag resistance movement against the faltering regime of Muammar Gaddafi, it was decided within NATO's North Atlantic Council that some sort of military intervention was needed. On 19 March, NATO commenced its Operation Unified Protector by launching Tomahawk missiles and air sorties at gov-ernment targets. The aims of the operation, set by the UNSC, included the establishment of a no-fly zone, the protection of civilians and the enforcement of an arms embargo. The unofficial aim, it was speculated, was regime change in favour of the National Transitional Council (NTC) – the Libyan resistance movement established during the war.

Operating in coordination with NTC but without ever deploying regular ground forces, NATO and coalition partners assisted in the gradual defeat of the Libyan government. Most of the support came from above, with aircrafts targeting vital government installations and its forces. The war raged until 20 October 2011, when during the battle of Sirte NTC forces located Qaddafi and beat him to death. Despite NTC requests that NATO stay on until the end of the year, the operation was formally ter-minated the following week. In the campaign's aftermath, NTC set up a new government, paved the way for elections and sought to establish and maintain a level of relative security.

The Western intervention in Libya in 2011 has been portrayed as a useful contrast to the costly and drawn-out campaigns in Iraq and Afghani-stan. For example, air power expert Christina Goulter (draft paper, p. 139) argues that 'after nearly a decade of counter-insurgency campaigns in Iraq and Afghanistan,... OUP proved that an air campaign, focused and driven by ISR [intelligence, surveillance, reconnaissance], can win a war when combined effectively with irregular ground forces'. Yet, in a powerful sense, the Libya campaign simply repeated the so-called Afghan Model, applied during the immediate combat phase of Operation Enduring Freedom and lauded then, too, as a uniquely effective means of applying Western military might (Biddle 2006). Then as now, the model saw Western powers ply their advanced combat capabilities – precision-guided munitions in particular – in support of a local ground force, reinforced by a small number of special operations forces to ensure proper coordination. Going back further, the prototype for the approach was tested in the Balkan campaigns of the 1990s, in which NATO aircraft bombed targets from a virtually risk-free altitude and let local allies (the Croat forces in Bosnia and the Kosovo Liberation Army in Kosovo) conduct ground operations.

The 'Libya model' presents undeniable advantages. First, the approach kept costs to a fraction of those accrued in Iraq and Afghanistan. Second, as in the NATO-led air campaign over Kosovo, coalition and civilian casualties were minimal; again, NATO intervened without incurring a single fatality. Third, although some ambiguity surrounded the actual aims in Libya, the results of the intervention appeared – at first blush at least – far more promising than those expected from Afghanistan following NATO's withdrawal.

These advantages notwithstanding, it is critical to acknowledge the preconditions that allowed the 'Libya model' to be at all effective. Indeed, the campaign was in many ways exceptional, undermining its status as a precedent. First, Col. Moammar Gaddafi's lack of subtlety, in combination with the backdrop of democratic revolutions in Northern Africa, provided the campaign with unprecedented international support – a sense of urgency to 'do something'. From then on, much of the war was fought in the desert, greatly facilitating aerial bombardment. There was also a clear opposition to Gaddafi in the NTC and the rebel troops that served as the proxy. Moreover, the geographic location, at the very borders of Europe, facilitated both basing and logistics. These conditions are not always going to present themselves in other contexts.

Going further, and risking a cliché, the enemy has a vote. Even in Libya, government forces sought to exploit NATO's strategic and tactical preferences. Having initially operated in large uniformed units across the desert, government forces adapted following the initial air attacks. As Brig. Ben Barry (2011, p. 6) explains, Gaddafi's forces dispersed heavy weapons in populated areas and made extensive use of armed 4×4 vehicles, similar to those used by the rebels', something that 'greatly complicated NATO's ability to identify and attack them'. Clearly, such adaptation came too late, yet future adversaries are likely to be more wily, severely limiting the viability of winning war from the skies.

Finally, it is worth considering the political consequences of the limited ownership inherent in this approach. The model inevitably empowers a local proxy. The key question, therefore, is what happens after the aerial bombardment has stopped, when the model is put back on the shelf and it is time to establish a new political accommodation that is both desirable and stable. These days, the Afghan war is hardly remembered for the initial successes of the 'Afghan model' – indeed it was precisely the political fall-out from the Taliban's toppling that has bedevilled subsequent efforts at stabilisation. Similarly, although successful in toppling the Gaddafi regime, the Libyan intervention unleashed destabilising forces within Libya and regionally. In Libya, 'factional, regional, tribal and ideological divisions' have marked the years since the revolution: the 'central government, far outgunned by powerful local militias, holds little sway beyond its offices' (*The Economist*, 22 February 2014). Regionally, fighters and weapons have spread as far as Mali and Syria, destabilising the already fragile states in the region.[1] The implication is not that NATO should have used ground troops in Libya, but rather that the Libya model must not be mistaken for more than it is: it does not render intervention easy, but simply offloads the responsibility for political consolidation onto others, with whom we must learn to work far more effectively.

The indirect approach

In recent years, the US has experienced a revolution in its understanding of counter-insurgency. When the US Army and Marine Corps published their counter-insurgency manual in December 2006, the term denoted, almost exclusively, the deployment of large armed formations to provide security for the host-nation population and assume responsibility for various military and civilian tasks. As the doctrine was written at a time of insurgency in Iraq in which 144,000 US troops were actively involved, this focus on the 'direct' approach to counter-insurgency was apposite. Yet even then, the manual was criticised for not acknowledging alternative approaches and this criticism has become far more vocal with the perceived failure of the direct approach in Afghanistan. The dominant argument now, and for the foreseeable future, is that for strategic, political and financial reasons outcomes must be achieved 'indirectly', by relying on the structures and capabilities of the host-nation partner and thereby doing more with less. While the following chapter will discuss military advising and assisting missions in more detail, this section highlights the most important opportunities, challenges and considerations of the indirect approach as an alternative to counter-insurgency operations.

A key precedent for this approach is the US advisory mission in El Salvador in the 1980s, which is credited with the defeat of the Farabundi Marti National Liberation Front. The British campaign in Dhofar, from 1962 to 1976, provides a second, increasingly cited precedent, since

Britain relied on the armed forces of the host-nation government along with sub-state militias to achieve its aims there. A more recent case is the US military's assisting of Colombia in its campaign against the FARC. This case provides the perfect foil for the direct interventions in Iraq and Afghanistan: they overlapped in time, but whereas the direct engagements were ruinously expensive, politically costly and ambiguous in their outcome, the weakening of FARC under President Alvaro Uribe is a counter-insurgency success story (Marks 2002). Similarly, the efforts led by the US special operations forces to assist the Philippines government against the Abu Sayaff Group stands out as a low-cost, low-profile yet fairly successful intervention, at least in comparative terms (Wilson 2006).

Proponents commonly point to five key advantages. First, the indirect approach puts local forces in the lead and thereby avoids many of the linguistic and cultural hurdles encountered by foreign troops. Second, by keeping the response local, the counter-insurgency campaign remains untarnished by the stigma of foreign occupation. Third, putting local forces in the lead also reduces the intervention's political costs for the intervening government. Fourth, these interventions are commonly less costly financially – a corollary of the smaller footprint (Luján 2013, p. 8). Fifth, and most fundamentally, the indirect approach puts the local government in charge of solving what is after all its problem: it predicates the solution on local ownership and responsibility.

The indirect approach rightly recognises the limits on what external powers can by themselves achieve in a foreign land, particularly one that they scarcely understand. The focus on partnerships also touches upon the essence of expeditionary counter-insurgency: the need to maintain host-nation legitimacy, build capacity and engage in a manner that is sustainable. Yet while the notion that 'small is beautiful' – that indirect deployments make more sense – is largely correct, it is deceiving and dangerous to stop the analysis at this point. Indeed, the indirect approach, like counter-insurgency or interventions of any type, involves severe challenges that must be fully understood. Three caveats stand out as critical.

First, experience indicates that working with and advising local security forces is an art all in itself. There is a common misconception that because the advisory approach puts the local government and its security forces in the lead, the intervening power is somehow shielded from the complexity otherwise typical of counter-insurgency. As is amply demonstrated by practice, advisory work is in fact highly challenging, requiring specific skill and capacity – something further discussed in the following chapter on military advising and assisting operations.

Two problems are historically consistent: ensuring the professionalisation of the host-nation security force; and ensuring that it uses what it learns in ways that are accountable and in keeping with mission objectives. In El Salvador, the congressionally mandated cap on deploying just 55 US advisors and the ban on them joining the El Salvadoran Armed

Forces (ESAF) on operation undermined the vied-for learning process. Specifically, US advisors lacked both leverage and oversight and relied, in other words, on ESAF being willing and able to follow the guidance provided. Neither of these conditions was obtained. Though the advisory campaign was vital for regime survival in the early phase of the war, the transition to peace a decade later had more to do with the passing of the Cold War and other domestic factors than the marvels of the indirect approach (Ucko 2013).

The problems of oversight and leverage resurfaced when U.S. troops sought to establish security forces in Iraq and Afghanistan. A consistent finding from these theatres, also echoed in Chapter 5, is that the effectiveness of advisory missions is best guaranteed by "partnering" with local security forces: living and operating with them, day and night, from the same base and streets. Yet the implications of this requirement are significant: it calls for specific and extensive preparation, including language training and cultural awareness. Brigadier Ian Gardiner (2006, p. 197), historian and veteran of the Dhofar campaign, illustrates this point well:

> The patience and tolerance to live harmoniously in an unfamiliar culture; the fortitude to be content with less than comfortable circumstances for prolonged periods; an understanding and sympathy for a foreign history and religion; a willingness to learn a new language; the flexibility and imagination and humility necessary to climb into the head of the people who live by a very different set of assumptions; none of these are to be found automatically in our modern developed Euro-Atlantic culture. These attributes, and the attitudes they imply, often have to be taught in addition to purely military skills.

Notwithstanding various efforts to boost regional expertise, it is uncertain whether U.S., UK or NATO troops are adequately prepared for this task. Pointing to the special operation forces as a solution, given their specialised skills, is insufficient: few in number and not easily mass-produced, they lack the capacity to undertake large-scale advisory missions – something that is further discussed in Chapter 6. To be sure, successful advisory efforts are rarely light in troop numbers: a mere 55 advisors may have deployed to El Salvador, yet El Salvador is a very small country, in close proximity to the United States where additional training was provided and, even then, the cap and other congressional restrictions undermined the proper prosecution of the campaign. To do better, sufficient advisors are required to accompany each unit being trained.

Therefore, the indirect approach cannot, must not, be seen as 'counter-insurgency on the cheap'. If partnering is indeed required, advisory missions will in all cases require sustained buy-in: institutionally to create the capabilities, and politically, to allow troops to operate from the front line over protracted periods, while embedded within host-nation forces taking fire. As seen in Afghanistan, it is often the advisors themselves who become the target, so as to sever the critical link that partnering provides.

Another consideration for the application of the indirect approach is the need for a partner. In Colombia, the Philippines, and most other settings where the indirect approach is said to have worked, the intervening advisors operated alongside an established government and military. Colombia, for example, has a long record of elected civilian governance and a strong military. By contrast, it is questionable whether the indirect approach would have worked in Afghanistan in 2001, in Iraq in 2003 or in other 'post-conflict' settings. This limitation clearly restricts the applicability of the approach.

Even where the central state is extant and somewhat competent, thorny issues of legitimacy and strategy still loom large. In the quest to defeat insurgency, the professionalisation of a country's armed forces or security sector is but one part of a broader puzzle. David Galula's (1964, p. 89) admonition that counterinsurgency is 80 per cent political and only 20 per cent military has become a cliché, yet have the statement's true implications been grasped? While having professional security forces is critical, it is not in itself strategically decisive: much depends on the *political objectives* that their operations serve. Where this strategy is misguided or altogether absent, security operations have little or no meaning. By analogy, it serves no purpose sharpening the scalpel if the surgeon operating is drunk.

This point is critical, as it is typically at the *political* level that the host-nation partnership will fray. Partners are more willing to accept military aid and assistance than undergo the political or social reforms also deemed necessary for success. Governments facing an insurgency almost by definition suffer from some legitimacy deficit – hence the organised armed resistance. It is not uncommon that they are more concerned with retaining power and privilege than with undercutting dissent through effective reform. The resultant dilemma for counter-insurgency advisors is formidable. In Dhofar, the solution to Said bin Taimur's refusal to reform was a military coup carried out by his own son and with the support of the British government. Within 24 hours, various liberalising measures were passed, giving political meaning to the armed forces' security operations and producing the happy outcome for which the campaign is known (Beckett 2008). Yet for a less happy precedent, consider the advisory years in Vietnam (1950–1965) and the US decision to remove the recalcitrant Ngo Dinh Diem, a desperate measure that opened the door to US ground troops in 1965. In other words, nothing within the indirect approach removes the need for suasion and compulsion – diplomatic tasks where the West underperforms. This once again limits what we can expect to achieve from the indirect approach: much like any other model of intervention, it must be tailored to specific circumstances and support a sound strategy.

Contingency operations

Another means of burden sharing is by delimiting the role of Western forces and ensuring that residual tasks are carried out by international, regional or local partners. The role played here might entail the provision of quick-reaction forces to assist a peace operation or protect it from a sudden crisis. Such a 'contingency operation' would in principle be similar to that played by the British military during its intervention in Sierra Leone in 2000 or by the French-led coalition force in Operation Artemis in eastern DRC in 2003. The benefit here is that, in coming to the assistance of a pre-existing mission, the intervening power is allowed to focus on just one phase of the campaign, thereby limiting its exposure and risk. Yet by the same token the effectiveness of these interventions also relies on the ability to transfer demanding follow-on responsibilities to competent actors with greater staying power.

Operation Artemis is a cautionary tale. In response to the destabilisation of eastern DRC, a French-led Interim Emergency Multinational Force (IEMF) deployed to Bunia to help shore up security and rescue the local UN peacekeeping mission. Per the conditions tied to its deployment, IEMF spent three months in Bunia, during which time it expelled militia elements and re-established security. It then handed over responsibility to the newly created UN 'Ituri Brigade', a 5,000-strong unit. On these merits, the operation was a success, yet the IEMF's limited mandate, temporally and geographically, meant that its effects were transient. As a later UN (2004, p. 14) report found, 'The strict insistence on the very limited area of operations – Bunia – merely pushed the problem of violent aggression against civilians beyond the environs of the town, where atrocities continued.' Moreover, despite the UN force's expansion, it remained undermanned and ill equipped to sustain the gains of the intervention, greatly undermining its longer-term significance (Berdal 2009, p. 112).

The British military has a more successful experience with 'contingency operations', one that illustrates the value of these types of interventions but also what they typically require. Initially deployed in Sierra Leone in 2000 to evacuate Westerners from the war-torn country, the in-country force commander, Gen. David Richards, saw an opportunity to side directly with the Freetown government against the Revolutionary United Front (RUF). British forces were involved in number of confrontations against the RUF and maintained a presence offshore to demonstrate resolve. The combat phase ended quickly but notably the British force then supported, trained and reinforced Sierra Leone's army and the local UN peacekeeping mission, so that the country's newfound stability could be sustained following the British withdrawal in 2002. Even then, Britain maintained a 140-strong force in Sierra Leone to advise the army and has remained one of the country's greatest bilateral donors of aid (Berdal 2009, p. 120).

While the results are far from incontestable, the point is clear: the effectiveness of military force depended on, inter alia, coordinated and properly resourced follow-up action. It calls for civil–military cooperation and the ability to raise the competence of local and international forces so as to enable a smooth transition. It is perhaps telling that many of the more successful contingency operations have been unilateral – thereby avoiding the extra coordination and timing challenges involved in multilateral campaigns. In that sense, the use of Western troops on contingency operations calls for many of the same capabilities as those needed for the indirect approach, which again highlights this area as requiring more urgent attention and prioritisation.

Conclusion – whither counter-insurgency?

Some argue that the limited success experienced with counter-insurgency so far in the twenty-first century is reason enough to reject it as a failed doctrine and to avoid this approach to operations in the future. Given the tremendous cost in blood and treasure, this is an understandable yet overly hasty and dangerous conclusion. While the international interest or willingness to 'fix failed states' may be seriously dented after the costly campaigns in Iraq and Afghanistan, it is equally true that most international military operations occur in environments where the state's reach and institutions have suffered significant damage: either the lack of state control is what causes intervention (Sierra Leone, Somalia), or it is the other way round, and the intervention results in the lack of a state (Afghanistan, Iraq). In either case, wherever Western armed forces will be operating, it is most likely to be in areas with weak formal structures of governance, where criminality, informal networks of patronage, the proliferation of small arms and sub-state politics are all common and need to be understood. In other words, in several types of expeditionary operations – even those that are not termed 'counter-insurgencies' – there will often be a need for the type of knowledge, skills and awareness called for and emphasised in counter-insurgency theory: how to engage with a civilian population, how to establish and maintain order, how to collect and process human intelligence, how to operate in foreign culture, how to provide basic services and so on. In Michael Howard's words,

> The military may protest that this is not the kind of war that they joined up to fight, and taxpayers that they see little return for their money. But ... this is the only war we are likely to get: it is also the only kind of peace. So let us have no illusions about it.
>
> (2007, p. 14)

The apparent inevitability of future counter-insurgency or counter-insurgency-like missions is clearly highly troubling. For most countries, it would imply an urgent need to learn from their past wars and to develop the required capabilities for nominally similar campaigns – to grapple seriously with the problem of insurgency and guerrilla warfare so as to guard against future upsets. While such a course of action is for many reasons advisable, it seems financially and politically improbable. So far, despite continual engagement in counter-insurgency campaigns for more than a decade, most countries have only taken a few steps institutionally in this direction. Over the longer term, such an endeavour would be further complicated by the constraints imposed by strained budgets, ever-shrinking ground forces, and the difficulty of assuring domestic support for costly and protracted operations. When the defence budget must compete fiercely with other sectors of public spending – health, education, infrastructure – it will be more difficult still to justify military adventures abroad of uncertain duration; indeed many already see these activities as creating more instability and terrorism at home than they will ever disrupt, or as distractions from higher-order priorities (Blagden 2009). On this last point, developing the capabilities required for counter-insurgency-type operations would also need to overcome the entrenched (conventional) prioritisation and cultures of defence ministries and military organisations and is therefore doubly unlikely. The most that may be hoped for is that countries with expeditionary ambitions maintain intellectual preparedness for these types of scenarios, even if they do not develop the specific and required capabilities. At the very least, the principles of counter-insurgency – and its focus on war as an intensely political endeavour – may serve as important balancing tools against the conventionally focused military cultures that tend otherwise to dominate.

Note

1 UN Security Council's Group of Experts, "Final report of the Panel of Experts established pursuant to resolution 1973 (2011) concerning Libya", S/2013/99*, 9 March 2013, 24–38.

References

Alderson, Alexander (2010). "The Army Brain", *RUSI Journal*, 155(3).

Barry, Ben (2011). "Libya's Lessons", *Survival*, 53(5).

Beckett, Ian F.W. (2008). "The British Counterinsurgency Campaign in Dhofar 1965–1975". In Daniel Marston and Carter Malkasian, Eds. *Counterinsurgency in Modern Warfare*. Oxford: Osprey.

Berdal, Mats (2009). *Building Peace After War*. Abingdon, Va: Routledge for International Institute for Strategic Studies.

Biddle, Stephen (2006). "Allies, Airpower, and Modern Warfare: The Afghan Model in Afghanistan and Iraq". *International Security* 30(3).

Blagden, David (2009). "Strategic Thinking for the Age of Austerity". *RUSI Journal* 154(6).

British Army (2009). Vol. 1 Part 10: *Counter Insurgency Operations*. Army Code 71749. London: Ministry of Defence.

Buley, Benjamin (2007). *The New American Way of War: Military Culture and the Political Utility of Force*. New York: Routledge.

Cassidy, Robert M. (2005). "The British Army and Counterinsurgency: the Salience of Military Culture". *Military Review* 85(3).

Cohen, Eliot, Conrad Crane, Jan Horvatch and John Nagl (2006). "Principles, Imperatives, and Paradoxes of Counterinsurgency". *Military Review* 86(2).

Cohen, Michael A. (2010). "The Myth of a Kinder, Gentler War". *World Policy Journal* 27(1).

Dixon, Paul (2009). " 'Hearts and Minds'? British Counter-insurgency from Malaya to Iraq". *Journal of Strategic Studies* 32(3).

Dunlap, Maj. Gen. Charles J., Jr (2006). "America's Asymmetric Advantage". *Armed Forces Journal*, September 2006.

Echevarria, Antulio J. II (2004). *Toward an American Way of War*. Carlisle, Pa: Strategic Studies Institute, US Army War College.

Egnell, Robert (2013). "A Western Insurgency in Afghanistan". *Joint Forces Quarterly* 70(3), 8–14.

French, David (2011). *The British Way in Counter-insurgency 1945–1967*. Oxford: Oxford University Press.

Galula, David (1964). *Counterinsurgency Warfare: Theory and Practice*. Westport, Conn: Praefer.

Gardiner, Ian (2006). *In the Service of the Sultan*. Barnsley, UK: Pen and Sword.

Gentile, Gian P. (2011). "COIN Is Dead: US Army Must Put Strategy Over Tactics". *World Politics Review*. www.worldpoliticsreview.com/articles/10731/coin-is-dead-u-s-army-must-put-strategy-over-tactics.

Goulter, Christina *Ellamy: The UK Air Power Contribution to Operation Unified Protector*. Draft paper in RAND study on Operation Unified Protector. Santa Monica, Calif.

Gumz, Jonathan E. (2009). "Reframing the Historical Problematic of Insurgency: How the Professional Military Literature Created a New History and Missed the Past". *Journal of Strategic Studies* 32(4).

Howard, Michael (2007). "A Long War?", *Survival* 48(4).

Kagan, Frederick W. (2006). *Finding the Target: The Transformation of American Military Policy*. New York: Encounter Books.

Kilcullen, David (2009). *The Accidental Guerilla: Fighting Small Wars in the Midst of a Big One*. New York: Oxford University Press.

Kiszely, John (2007). *Post-modern Challenges for the Modern Warrior*. Cranfield, UK: Defence Academy.

Kitson, Frank (1977). *Bunch of Five*. London: Faber and Faber.

Kitson, Frank (1971). *Low Intensity Operations: Subversion, Insurgency, Peace-keeping*. London: Frank Cass.

Luján, Fernando (2013). "Light Footprints: The Future of American Military Intervention". *Voices from the Field*, Center for a New American Security, March 2013.

Luttwak, Edward N. (2007). "Dead End: Counterinsurgency Warfare as Military Malpractice". *Harpers Magazine*, February 2007.

Mackinlay, John (1998). "War Lords". *RUSI Journal* 143(2).

Marks, Thomas (2002). *Colombian Army Adaptation to FARC*. Carlisle, Pa: The Strategic Studies Institute. Marks, Thomas A. (2009). "Mao Tse-Tung and the Search for 21st Century Counterinsurgency". *CTC Sentinel* 2(10) (October), 17–20.

Mockaitis, Thomas R. (1995). *British Counterinsurgency in the Post-imperial Era.* Manchester, UK: Manchester University Press.

Nardulli, B., K. Pollack, T. Szayna and B. Watts (1999). "Coup-proofing James T. Quinlivan". *International Security* 24(2), 131–165.

Quinlivan, James T. (1995). "Force Requirements in Stability Operations". *Parameters* 25, 59–69.

Quinlivan, James T. (2003). "Burden of Victory: The Painful Arithmetic of Stability Operations". *Rand Review* 27(2).

Rigden, I. A. (2008). *The British Approach to Counter-insurgency: Myths, Realities, and Strategic Challenges.* Available from: US Army War College.

Smith, Gen. Rupert (2005). *The Utility of Force: The Art of War in the Modern World.* New York: Knopf.

Thompson, Robert (1966). *Defeating Communist Insurgency: The Lessons Of Malaya and Vietnam.* New York: Praeger.

Ucko, David H. and Robert Egnell (2013). *Counterinsurgency in Crisis: Britain and the Challenges of Modern Warfare.* New York: Columbia University Press.

Ucko, David H. (2013). "Beyond Clear-Hold-Build: Rethinking Local-Level Counterinsurgency after Afghanistan". *Contemporary Security Policy* 34(3).

UN Peacekeeping Best Practices Unit (Military Division) (2004). *Operation Artemis: the Lessons of the Interim Emergency Multinational Force.* Available from: UN Department for Peacekeeping Operations.

UN Security Council's Group of Experts (2013). *Final report of the Panel of Experts established pursuant to resolution 1973 (2011) concerning Libya.* Available from: UN Security Council.

US Department of the Army and United States Marine Corps (2006). *Counterinsurgency.* Available from: US Department of the Army.

Weigley, Russell F. (1973). *The American Way of War: A History of United States Military Strategy and Policy.* Bloomington, Ind: Indiana University Press.

Wilson, Gregory (2006). "Anatomy of a Successful COIN Operation: OEF-Philippines and the Indirect Approach". *Military Review* 86(6), 2–12.

Woodward, Susan L. (2007). "Do the Root Causes of Civil War Matter? On Using Knowledge to Improve Peacebuilding Interventions". *Journal of Intervention and Statebuilding* 1(2).

5 Military advising and assistance operations

Guro Lien

This chapter discusses military advising and assistance operations, with an emphasis on the challenges and opportunities for these types of operations in the future. The increasing use in operations of military advising and assistance forces, whose main task is to train and mentor indigenous security forces, is mainly driven by two of the trends described previously in this book. First, the financial troubles in Europe have led to decreasing defence budgets and cuts in capabilities in many Western states. Military advising and assistance operations can be a lot less resource demanding, in terms of personnel, financial costs and equipment, which makes them an attractive option for nations with decreasing defence budgets. Second, the decreasing recruitment base for Western militaries due to changing demographics may make it difficult to sustain large deployments overseas in the future. Also, technological advances in the future may make military operations more efficient, but also more costly. The trend towards ever smaller military forces in the West will continue, and thus the capacity of local security forces to handle internal security is all the more important. A military advising and assistance operation demands fewer troops over a shorter time span, and will reduce the burden on international forces by enabling indigenous security forces to provide security. On the other hand, the requirements of advisors, in terms of maturity, personal qualities and education, are different from those of combat soldiers. This may entail an increasing deployment of higher-ranking and more mature officers, of which there are few available.

A further reason for the increasing emphasis on military advising and assistance operations has been the search for exit strategies in complex military interventions. Peace and stabilisation operations have become very complex and often long-lasting, and activities like reconstruction, reintegrating former combatants, supporting democratic development and strengthening the rule of law are considered part of the responsibility of the intervening forces (Luck 2006, p. 33). These activities cannot be carried out by military forces alone but require a system-wide institutional approach with an emphasis on building local capacity. However, building state institutions is very complicated and time consuming, and in the meantime, sustaining a foreign military presence becomes increasingly difficult. As stated in the Brahimi report from 2000, 'while the peacebuilders may not be able to function without the peacekeepers' support, the peacekeepers have no exit without the peacebuilders' work' (Luck 2006, pp. 40–41). Rupert Smith has also acknowledged this dilemma. He writes that Western states today 'do not intervene in order to take or hold territory; in fact, once an intervention has occurred a main preoccupation is how to leave the territory rather than keep it' (Smith 2005, p. 270). Building local capacity, so that the long-term responsibility for peace and development can lie in the hands of the local security forces and ministries, becomes an increasingly attractive option as an exit strategy.

This chapter starts with a brief overview of the history of military advising and assistance missions. The next part of the chapter discusses some recent experiences with military advising and assistance and what future developments we may encounter when it comes to military advising and assistance. This chapter only discusses military advising and assistance operations as performed by regular military units, and not by special operations forces, although some of the points made may be relevant for military assistance performed by special operations forces as well. That specific kind of military assistance is discussed more by Johansen in Chapter 6. Also, the more strategic aspects of security sector reform and security force assistance is dealt with in Chapter 4 by Egnell and Ucko, and will not be dealt with specifically here.

Military advising in a historical context

Western states providing military advising and assistance is not a recent phenomenon. In fact, there are examples of such missions from the 1800s. The most famous example is undoubtedly T. E. Lawrence, also known as Lawrence of Arabia, who is known especially for his liaison role during the Arab Revolt against Ottoman Turkish rule of 1916–1918. However, this was not the first time Egypt received military assistance from Europe. One of the earliest examples of a formal military advising mission is the French mission to Egypt in 1815. The mission was largely composed of veterans from the Napoleonic Wars and was assigned with modernising and institutionalising the Egyptian Army along Western lines (Stoker 2008, p. 2). Chile, after gaining independence from Spain in 1818, also hired German and French military advisors to help build its new army. Interestingly, Chilean officers later provided military advisor teams to other Latin-American countries (Stoker 2008, p. 2). These early missions were often used as tools to increase economic influence, seeking to further trade relations and fostering a market for arms and other military equipment abroad.

After World War II, military advising became a tool to further ideological aims. The US committed advisory missions to Korea, Vietnam and

China, and the Soviet Union invested heavily in foreign advising of Cuba and Egypt. According to Nilsson and Zetterlund, 'the objective of the support was to form well-equipped and well-trained security forces that could participate in the ideological and political struggle both domestically and internationally' (2011, p. 36).

The most studied example of military advising missions is probably the US involvement in South Vietnam before and during the Vietnam War, the longest and largest military advising effort of the US to date. The first advisors arrived in Saigon in 1950, as part of the United States Military Assistance Advisory Group, Vietnam. Originally, the mission was tasked with supervising the donation of $10 million-worth of military equipment to support French legionnaires in their fight against the Vietminh forces, but the scope and role of the advisor group quickly expanded. Up until 1960, the US had between 750 and 1,500 military advisors in Vietnam, helping to train the South Vietnamese Army. By 1961, after the Kennedy administration decided to increase the effort, about 3,400 US advisors were engaged in Vietnam, and by 1963 the number had grown to over 15,000 troops (Miller 2013). The advisors and the combat troops withdrew from Vietnam only in 1973, after the signing of the Paris Peace Accords.

The effort in Vietnam has largely been considered a failure, due to inadequate attention to the selection and training of advisors, the increasing politicisation of the Vietnamese Army (with the consequence that military leaders were chosen and promoted based on political loyalty, not on military professionalism), and the tendency to organise, equip and train the Vietnamese army based on US standards and procedures (Ramsey 2006a, p. 110). Interestingly, many of the same problems are identified in a recent study of the military advising effort in Afghanistan (Kelly *et al.* 2011).

Although different in terms of scope and motivation, these early missions have two things in common. First, it was generally the great powers that were requested to provide assistance to smaller nations (Stoker 2008). The US is still the largest provider of military advisors today, but increasingly smaller nations also contribute with these types of missions in international military operations. Second, the early missions focused mainly on technical support and developing the military operational capabilities of the host nation (Nilsson and Zetterlund 2011). Today, it is widely recognised that the institutional and organisational capabilities of the security sector must also be taken into account, by supporting and advising ministries and general staffs (Dubik 2009; Kelly *et al.* 2011).

Present trends and future trajectories – doing more with less?

As economies contract and militaries shrink, smaller and shorter deployments may become increasingly popular in the future. This may mean that demand for military advising and assistance missions will increase. Building local security capacity was once the domain of special operations forces, but with increasing demand it may also become a core assignment for conventional forces. It could also mean an even greater reliance on private military firms to perform these functions. This section sums up some of the trends already apparent in today's military advising and assistance operations, and what implications they may have for the future. Because of the very people-centric nature of military advising and assistance, with its emphasis on personal relationships and inter-personal skills, the majority of this section will discuss the parameter of relative force composition and strength. Military advising and assistance operations may take place across all operational environments, conflict intensity may vary, consent of the host nation government is normally a given and a mandate is not always necessary for these types of operations. Therefore, those parameters will be dealt with more briefly. The most recent experiences with military advising and assistance operations have been as exit strategies in Iraq and Afghanistan, as discussed next.

Mentoring and training missions as exit strategies

Several of the most recent military interventions have started out as so-called Counter-Regime operations (Norheim-Martinsen *et al.* 2011). In these operations, military force is used to swiftly oust an incumbent regime from power, using massive military force. This type of operation is often successful, in that it achieves the operational objective that is the aim. However, removing a dictator or stopping an invading force is only part of the solution. According to Rupert Smith, the primary role of the military in today's interventions is to create a 'condition', or, in other words, to establish security, so that other instruments of power can tackle the issues of governance and development (Smith 2005, p. 270). But building state institutions is very complex and challenging and may take decades, and, as pointed out by Kjeksrud in Chapter 7, there is a growing sense of state-building fatigue. In the meantime, a foreign military presence can become difficult to sustain, both economically and politically. Building local capacity, so that the long-term responsibility for peace and development can lie in the hands of the local security forces and ministries, becomes an increasingly attractive option as an exit strategy (Nilsson and Zetterlund 2011, p. 45). The most recent examples of military advising and assistance used as exit strategies have been Iraq and Afghanistan.

In Iraq, there has been considerable effort to train the Iraqi army since 2005, mainly by the US Military Transition Teams (Stoker 2008). In addition, a NATO-led training mission operated in Iraq from 2004 to 2011, helping to train Iraqi military and police forces (NATO Official Homepage 2012). The combat mission in Iraq formally ended on 1 September 2010, and from then on until the US forces withdrew in 2011 the 50,000 remaining US troops focused on advising and assisting Iraqi security forces (Cooper and Stolberg 2010). The responsibility for security in Iraq now lies in the hands of local Iraqi forces. During the NATO Summit in Chicago in May 2012, an exit strategy for Afghanistan was endorsed, ensuring a withdrawal of the majority of foreign troops by the end of 2014. But this does not mean an end to foreign presence in Afghanistan. Instead of regular combat troops, there will be an increase in the military advising and assistance operations. The US has had Embedded Training Teams in Afghanistan since 2003 to mentor and advise the new Afghan National Army. The NATO-led Operational Mentoring and Liaison Teams (OMLTs) first deployed in 2006 and consist of multinational teams of military advisors that follow Afghan companies and battalions in both training and combat operations (Kelly *et al.* 2011). Since mid-2013, all military operations have been led by Afghan troops, with coalition forces as partnering units.

The experiences of military advising and assistance in Iraq and Afghanistan, where Western nations have found themselves tasked with building military forces, police forces and ministerial capacity almost from scratch, may suggest that Western nations will avoid such missions in the future. The task in Afghanistan has been especially complicated because the insurgency increased the risks and instability in the country, making it all the more difficult to build capacity and train forces while they were fighting at the same time (Kelly *et al.* 2011; Stoker 2008). However, although the desire to avoid 'a new Afghanistan' is strong, that is not to say that Western nations will not be conducting military advising and assistance missions in the future. After all, states have been sending military advising units to other countries since the 1800s, and it is unlikely that they will cease completely. But the nature of military advising and assistance operations may go back to being more limited in scope and ambition. Previously, military advisors largely dealt with existing forces (Stoker 2008, p. 233). Supporting security forces in the making is less strenuous, and with a strong and capable central government, the advisors can focus on combat training. Building and reforming security institutions such as ministries is very sensitive and can be controversial. Thus we may see a move away from state-building tasks in the future and more towards providing traditional combat training and advice. According to Nilsson and Zetterlund, the experiences of Western states in training security forces in tactics and strategy have been more successful than the efforts to promote good governance and democracy (Nilsson and Zetterlund 2011, p. 36).

Relative force composition and strength

In recent years, defence budgets have been cut in almost all countries, and the financial crisis has led to an increasing emphasis on more limited international military engagement. Mentoring and training operations can be less resource demanding in terms of personnel. Teams of military advisors are smaller units that can function as force multipliers for indigenous troops. They therefore make attractive options for smaller nations who want to contribute internationally but do not have the resources to contribute large troops or sophisticated weapons systems. The EU has increased its use of mentoring and advisors in recent years, in part because they are cheaper to sustain and less controversial. In parallel with a training mission in Somalia, the EU has deployed 175 officers to the Horn of Africa, to train and advise indigenous officers in Djibouti, Kenya and the Seychelles. More recently, the EU deployed a training mission to Mali to help train the Malian Armed Forces and reform the security sector in that country.

An increase in the demand for military advising operations could have consequences for the structure of Western military forces in the future. It may mean that regular forces must perform advising and assistance tasks to an even greater extent, and this will in turn require more specialised training and preparation for these types of deployments. A review of various studies and lessons learned from mentoring missions showed that all emphasise the need for specialised training for these types of missions (Kelly *et al.* 2011; Haug 2009; Stoker 2008). The experiences of previous mentors indicate that traditional soldier training may not be sufficient or appropriate for military advising and assistance operations. According to Kelly et al., there is 'little in a normal career [that] prepares a soldier for advising other than previous advisory assignments' (Kelly *et al.* 2011, p. 82). In addition, not all soldiers are suited 'by character and personality for the advisory role', and thus selection criteria for advising missions must take these issues into account (Kelly *et al.*, p. 83). There is a greater need for cultural awareness and language skills, and less adherence to hierarchy and rank. Smaller units of advisors in close contact with foreign soldiers indicate that interpersonal skills become more important.

If we are to engage more in these operations in the future, we may have to change our approach towards planning and educating soldiers for these types of missions. Some countries may choose to set up a permanent roster of previous and willing advisors, including both military and civilian staff. An even more drastic approach would be to develop niche capacities specialising in advising and assistance. If these types of missions increase, and there is a general development towards making them part of an exit strategy for all new operations, we may even see countries establishing permanent units of advisors. For a smaller nation, which may have difficulty deploying larger numbers of combat troops, this could be a way of contributing to operations and being a force enabler through supplying a niche capacity. For larger nations, a smaller dedicated unit with specialised advisors may make the overall operation more effective and reduce the risk of mission creep. There have in fact been discussions in the UK about the possibilities of forming permanent capabilities for military assistance, security and development tasks (Haug 2009, p. 19).

In the most recent advisory missions, military forces have to a large extent been tasked with training police forces. This is not something military soldiers are trained to do, but 'in large-scale cases there will likely be no other organisation that can take on this mission' (Kelly *et al.* 2011, p. xx). However, it runs the risk of turning the police forces into paramilitaries instead of law enforcement agencies. We may therefore also see an increase in the development of permanent expeditionary civilian police units in the future, specialising in advising and mentoring new police forces. The UN has a standing policing capacity, but these forces numbered only 41 officers in 2010 (Durch and England 2010). There is also a European Gendarmerie Force (EGF), comprised of police officers from France, Italy, the Netherlands, Portugal and Spain, who specialise in crisis management but have also contributed to training the Afghan National Police under the International Security Assistance Force (ISAF). These forces have military status, however, and conduct mainly militarised police functions. The EGF consists of up to 800 officers, and can be put at the disposal of the EU (first and foremost), the UN, OSCE, NATO, other international organisations or ad hoc coalitions. But the lack of more civilian deployable police capacity has been lamented in the most recent training and mentoring missions.

As smaller nations with decreasing budgets and troop numbers are increasingly called upon to assist other small nations, we may see an increase in multinational military advising and assistance operations in the future, as already seen in the OMLTs in Afghanistan. This brings many challenges, such as interoperability of equipment, language problems and differences in doctrine and standards. In addition, national caveats, which limit what some officers can and will do in operations, have been an impediment to the efficiency of the OMLTs in Afghanistan (Kelly *et al.* 2011). Norwegian officers in the OMLT have stressed the importance of common training and education, preferably with OMLT officers from other countries (Krekvik 2011). When the hand-over/take-over period in Afghanistan was simultaneous for the nations contributing to the same OMLT, the mentoring effort greatly improved (Krekvik 2011). In the future, if this type of mentoring effort is continued in other countries, we may see bilateral agreements to offer mentoring and training teams composed of officers from more than one nation. If the teams are trained together and have clear guidelines as to what doctrines and standards to teach, many of the challenges could be overcome. Or, conversely, because of the many challenges in conducting military advising and assistance in multinational teams, we may see a decreasing interest in these types of missions. Multinational military advising and assistance operations bring little prestige to the contributing nations. Sophisticated weapons systems, such as fighter planes, bring a lot more attention from the media, from international organisations and among politicians. Taken together, this may lead to smaller nations developing niche capabilities that bring more prestige, rather than focusing on training and advising. Perhaps the greater powers will have to bear the main responsibility for these types of missions in the future.

In addition, and as already witnessed today, we might see a development towards increasing 'south–south' cooperation, where Western-trained soldiers in developing countries take on advising missions in other developing countries. For instance, police officers from Bosnia have been part of a UN mission in South Sudan, strengthening the local police forces and participating in police training. The EU Police Mission in Bosnia, in which police officers from the EU trained Bosnian police officers, ended in June 2012. Another example is from a Security Sector Reform effort in South Sudan, where government employees from Kenya mentored and advised civil servants in South Sudan (Tarp and Rosén 2012). This type of cooperation could be beneficial, in that the advisors have experience in being advised, as well as, in some cases, having similar force structures, equipment and ambitions. Being geographically closer may also make the effort more sustainable in the long run, and it is seen as more effective because of the 'cultural and linguistic similarities, similar value and administrative systems, and knowledge of local and regional conditions' (Tarp and Rosén 2012, p. 16). Western support for south–south cooperation could become increasingly important in the future. It would also lessen the burden on Western defence budgets.

A relatively new development when it comes to military advising and assistance operations is the increasing role of private security companies. One of the first examples of this was the Military Professional Resources Incorporated (MPRI) in Croatia in the 1990s (Singer 2008). The American-based company provided extensive training and advanced military planning assistance to the Croat Army, and some have claimed that they were instrumental in the success of the Croat army in fighting the Serbs during Operation Storm in 1995 (Singer 2008, p. 5). MPRI is one of the largest private companies providing military training and their 15,000-employee roster is largely retired military personnel (Stoker 2008, p. 244). More recently, MPRI was hired to assist the Americans in reforming the Afghan Ministry of Defence (Hodes and Cedra 2007). The operation in Iraq saw a large number of private security companies, and some have even called the operation a 'coalition of the billing', with more than 100,000 private military contractors deployed (Stoker 2008, p. 224; Singer 2008). The training and reform of the Iraqi security sector was but one aspect that was outsourced to private firms. Iraq was also the first time private companies were tasked with developing and building a new security sector; previously they had been hired to train already existing forces (Stoker 2008). Using private companies is less costly, and there is usually less media attention in case of casualties. It also reduces the burden on experienced and well-trained military officers, who may be needed at home or in other operations. However, the lack of regulation means that there is no control over 'who can work for these firms and who these firms can work for' (Singer 2008, p. 255).

An increase in the demand for military advising operations may also lead to an increased reliance on private security firms to perform these tasks in the future, because regular forces may become overstretched. There is very little regulation of private military companies, and, according to Singer, they 'exist within a gray area of the law, with an uncertain legal status and minimal accountability' (Singer 2008, p. 251). A continued development in this direction may force a reining in of the power of these companies. We might see more regulation and demands that private firms be held accountable also. There have been debates in the US over whether some functions should not be outsourced in the future, or whether contracted soldiers should be placed under the US military's Uniform Code of Military Justice (Singer 2008, p. 258), meaning they can be court-martialled. The EU has already taken steps in this direction, and recognised the importance of regulating military services such as technical assistance, maintenance and training (Krahmann 2005). The combination and harmonisation of national and international legislation has converged in the EU and has, according to Krahmann, contributed to the strengthening of controls over these companies. An increased reliance on private military companies may also entail increasing regulation of these actors in the future.

Conflict intensity

There is a misconception that military advising and assistance operations are low-risk operations and mainly involve training inside a camp. Although the Army Field Manual for Security Force Assistance recognises that military advising can be undertaken across the entire spectrum of conflict, it nonetheless states that 'in most situations involving this assistance, there is relatively little weight on offensive and defensive operations' (US Army Field Manual 2009 3–07.1, 1–1). But as Owen West has described in his book *Snake Eaters*, 'combat training is combat', and it can entail high conflict intensity and actual combat situations (West 2012, p. xx). The most recent examples of military advising and assistance operations show that, more typically, conflict intensity varies, depending on what the unit is training for (Krekvik 2011). This can be very stressful for the individual soldier, because their situation seems unpredictable and tenuous. Another risk factor in recent years has been so-called green-on-blue attacks, where American or coalition soldiers are attacked by their local partners. This may be caused by infiltration by insurgents or rogue elements, or it may be caused by misunderstandings and misperceptions. In Afghanistan, these types of attacks have increased since 2009 and accounted for about 14 per cent of coalition deaths in 2012 with more than 50 Western officers killed (Gohel 2012). Increasing infiltration by

Taliban or other insurgent groups greatly increases the risk for advisors, and may cause stress and anxiety.

Employing troops in mentoring and training operations is similar to the main tasks that the military are trained to perform, and it is also a mission that military officers find meaningful. It is, according to Rupert Smith, a way of finding utility for military force in peacekeeping operations (Smith 2005, p. 320). A study of Norwegian mentors in Afghanistan showed that the officers found the tasks recognisable – they were in a high-intensity combat situation with typical military tasks that they were trained for (Lien 2014). In addition, their roles as military officers were clear, compared with the more undefined humanitarian tasks in low-intensity peacekeeping operations. But although military officers are trained to deal with high conflict intensity, these types of missions offer new challenges in that mentoring officers are always, in a way, 'on duty'. When they live, eat and sleep with their local counterparts, officers may experience a certain degree of conflict intensity at all times. Thus, it can be very demanding for the individual officer to serve as an advisor to a foreign military. Several studies claim that mentoring and advising should only be undertaken by experienced and mature officers, preferably with experience in mentoring. It is a difficult task, and it requires interpersonal skills, cultural sensitivity and, according to John H. Cushman, who served as a military advisor in South Vietnam in 1972, an 'unmilitary philosophical or reflective bent' (in Ramsey 2006b, p. 51). As explained by David L. Shelton, a former advisor of the US Army serving in El Salvador in the 1990s (in Ramsey 2006b, p. 65):

> every advisor is placed in the difficult position of trying to influence the behavior of others over whom he has no authority, causing them to do things that may be foreign to their nature and habit, while at the same time attempting to interpret, implement, and respond to criticisms of US political decisions over which he has no input or control.

In addition, advisors often encounter ethical problems such as corruption (Kelly *et al.* 2011; Krekvik 2011). This creates difficult dilemmas and moral anguish for the advisors, who are part of a reform effort that will lose legitimacy if corruption is widespread. The advisors must maintain good working relations with their counterparts, which may be damaged if they address the issue directly. Another ethical dilemma arises because of national caveats (Kelly *et al.* 2011). National rules may dictate that advisors cannot go on certain types of missions or in certain areas, which causes problems when the local unit is planning operations that violate national regulations. The advisor faces the difficult choice between supporting his partners and following national regulations. This may cause stress for the Western advisors, who face difficult ethical choices on a regular basis. There may also be moral issues related to the conduct of indigenous forces when it comes to the treatment of civilians and prisoners. Owen West claims that prisoner abuse was common in the Iraqi army, even when they were under American guidance. In his book, he describes a situation where one of the advisors had been badly injured by a roadside bomb, and the Iraqi troops have imprisoned a suspect for planting the bomb (West 2012, pp. 136–137). The Iraqi Major implies that by using rougher interrogation methods they can get the information they need to get a conviction, but the American advisor knows that this is against protocol. The advisor faces the dilemma of punishing the man that hurt his colleague by turning a blind eye or following the rules of engagement and risking letting the man go free to plant other bombs in the future. These types of dilemma can be difficult and frustrating for officers and cause a great deal of stress.

It is difficult to envision great changes when it comes to conflict intensity in future military advising and assistance operations. Recent history has shown that conflict intensity varies throughout the operation and that the moral anguish and perceived conflict intensity for the mentors can be high although combat intensity may be low.

Operational environment

Mentoring and advising at times take place in harsh conditions in unfamiliar environments, and the operational environment and facilities available can be challenging for Western soldiers used to certain standards. For instance, mentoring efforts since 2000 have taken place in Somalia, the Democratic Republic of Congo, Sudan, Uganda, Afghanistan and Iraq, all underdeveloped countries. Especially in advisory missions, where mentors live and train with their local counterparts, living standards may be different from what the soldiers are used to. However, the Norwegian mentors to the Afghan National Army felt that these issues were exaggerated and that it did not cause problems to live closely with their Afghan counterparts (Krekvik 2011, p. 41). On the other hand, the Western soldiers are often better equipped than indigenous soldiers, especially when it comes to personal equipment such as clothing and shoes.

In military advising and assistance operations, the operational environment also comprises the local officers and soldiers that mentors are operating with and among, the so-called human terrain as described in Chapter 2. Studies have shown that the main frustrations for advisors in the field are usually cultural issues rather than the environment or lack of facilities (Krekvik 2011; Kelly *et al.* 2011; Ramsey 2006b). In Afghanistan and Iraq, the idea of 'Inshallah', or 'God willing', has been difficult to grasp for Western advisors (Krekvik 2011; Stoker 2008). Advisors from the US Army in Iraq have reported that the soldiers they were training would sometimes refrain from seeking cover while being shot at by insurgents because, 'Inshallah', they would survive if God wanted them to (Stoker 2008).

There have also been examples where Iraqi recruits have closed their eyes while shooting because, 'Inshallah', they will hit the target if God wants them to (Stoker 2008). Similarly, an American advisor to the Saudi National Guard in 2005 encountered problems due to hierarchy and age because power traditionally derives from being older in Arab culture (Ramsey 2006b, p. 97). In the US Army, rank gives you the power to make decisions, but this is not the case in the Arab world. The American advisor spent several months trying to understand why a senior officer would not make even simple decisions but eventually discovered that the senior officer did not, according to Arab culture, have any real power to make decisions in the first place (Ramsey 2006b, p. 97). US advisors in Vietnam faced several cultural challenges. The Americans were highly systems-oriented, and used a direct approach to problem solving. But, according to Ramsey (2006a, p. 49):

> the Vietnamese ... did not see fitting themselves into systems and patterns as all that important. It was the commander, with his knowledge, courage, good instincts and good luck, with possibly the Mandate of Heaven thrown in, who was all all-important. Systems were things to be either used or circumvented, for the benefit of the commander or his family, or perhaps his units, in some cases.

Consent

Local capacity building almost always means supporting the government in power, in both training its armed forces, and building ministerial capacity. Therefore, the consent of the government is a given. However, in some cases it may be desirable to support elements opposing the government to force a regime change, for instance as seen in Libya in the fight against Gaddafi. But this specific type of military assistance is rarely, if ever, undertaken with regular security forces today. It is more common to deploy special operations forces to perform such tasks, as described in Johansen's chapter. However, as discussed in Chapter 4 by Egnell and Ucko, in an ethnically fragmented society, the government may not be the main provider of security. Local ownership and consent from the population is complex and difficult, and some researchers have used the term 'selective' or 'factional' ownership instead (Giustozzi 2008). The security sector reform agenda, in particular, is state-centric and builds on the Western model and principles of 'good governance' (Nilsson and Zetterlund 2011). But in many post-conflict countries, the government may not be the main provider of security. In fact, non-state or sub-state actors may have a responsibility for providing security, whether or not they are linked to the state (Nilsson and Zetterlund 2011, p. 31).

Also, there is a growing recognition that there may be cases where supporting a non-state actor with capacity building may be beneficial in countries where the central state is weak. For instance, Afghanistan has never had a strong centralised state but has relied more on local actors to provide security (Nilsson and Zetterlund 2011, p. 29). This may make it difficult to implement a top-down approach to security sector reform and military advising. This trend may increase in the future. In fact, such an approach was initially tried out in Afghanistan in 2003, when the US decided to train militia leaders as company-grade officers with local warlords as generals (Younossi *et al.* 2008, p. 14). However, this effort was quickly abandoned due to political conflicts and lack of progress. In the future, we may find Western states attempting to support non-state actors, albeit on a limited scale. In line with wanting to avoid grand state- and nation-building endeavours, we might see more limited efforts at building capacity in specific geographical areas or among specific population groups, in part because it would be less resource demanding.

A number of scholars warn against such non-state or sub-state capacity building (Martens 2012; Berman 2010). Supporting one faction may breed resentment among those who are not favoured, and this will in all likelihood lead to increasing violence and instability. While supporting local tribal leaders or warlords may provide short-term security, it rarely leads to long-term stability (Martens 2012, p. 185). Martens warns that supporting warlords can lead to 'a static situation where no-one who has any power has an incentive to replace warlordism with institution building' (p. 200). It may end up undermining the legitimacy of the state and, in the long run, fuel the conflict by altering the balances of power (p. 30). However, it must be taken into account that the chances of success are small, and that supporting sub-state actors may in fact fuel conflicts instead of resolving them. For these reasons, supporting non-state actors is not endorsed by the UN, and it would be very difficult to include this type of capacity building strategy in an operational mandate (Nilsson and Zetterlund 2011, p. 31).

Mandate

A mandate for military advising and assistance operations may not always be necessary. Especially in operations where the host nation requests military assistance, a mandate is often redundant. And if a mandate were deemed a necessity for these types of operation, it would normally be quite easy to acquire because the host nation has agreed to the operation. However, there may be situations where other UN members oppose such an operation, because it is seen as meddling in the internal affairs of another state. Military assistance and security sector reform are at the very heart of the state- and nation-building agenda, and ethnic and regional sensitivities may come into play. Thus, even in cases with host nation support for military advising and assistance operations, discussions surrounding a mandate may arise in the future.

International operations in the future are likely to be of smaller size and cost, and the desire to avoid 'a new Afghanistan', with a lengthy and complicated state-building enterprise, is strong. This may mean that, in the future, an exit strategy in the form of local capacity building of the security sector will be an integral part of most mandates at the outset of the operation. This is already apparent today, and may only increase in relevance in the future. The search for exit strategies has increased the importance of security force assistance and security sector reform and this trend is likely to continue in the future. So even though acquiring a mandate for military advising and assistance operations is less pertinent than for military interventions, it may still be relevant for these types of operation also.

Concluding remarks

With decreasing economies and smaller militaries, we may see an increasing interest in smaller and shorter deployments in the future. This may mean that demand for military advising and assistance missions will grow, and it may also entail military advising becoming a core assignment for conventional forces. Military advising and assistance operations are one way of bridging the gap between intervention and independence by enabling local security forces to provide protection and security for the population. But there is also a growing recognition that the institutional and organisational capabilities of the security sector must be taken into account in order to ensure more long-lasting stability and peace. Security and development are two sides of the same coin. Lasting security cannot be achieved without providing for the basic needs of the population, while development is impossible without security. This means that security cannot be created by military means alone – it requires good governance, justice and the rule of law, as well as reconstruction and development. However, state building is an extremely complicated business. In the West, there is a sense of 'war fatigue' after lengthy engagements in Afghanistan and Iraq, and there is a general reluctance to take on complex peace and stabilisation operations in the near future. These two operations have also highlighted the inherent difficulties in building states from scratch based on Western models and standards.

The purpose of this chapter has been to open up the debate about military advising and assistance operations in the future by discussing freely a number of possible future scenarios. A number of developments may occur, such as an increased emphasis on developing deployable civilian police capacities or increased regulation of private military firms. We may also see more limited efforts in military advising and assistance in the future, with a move away from ambitious state-building projects and perhaps an increased support for south–south cooperation. From a historical perspective, it is interesting to note that many of the same problems concerning the education, selection and training of advisors seen from the Vietnam effort are still reported from more recent operations. In addition, the main frustration for soldiers both today and previously has been cultural issues. Perhaps this insight will prompt some states to focus more on specialised education and training for advisors, as well as a more rigorous selection process. We may also see the development of niche capacities in this area, but traditionally military advising has brought little prestige, and therefore states may choose to concentrate on more high-tech sophisticated weapons systems instead. Whatever the future holds, military advising and assistance has been an element of the military profession since the 1800s, and it will continue to be in the future.

References

Berman, Sheri (2010). "From the Sun King to Karzai: Lessons for State Building in Afghanistan". *Foreign Affairs.* 8(2), 2–9.

Cooper, Helene and Sheryl Gay Stolberg (2010). "Obama Declares an End to Combat Mission in Iraq". *New York Times* online version, 31 August 2010. Retrieved from www.nytimes.com/2010/09/01/world/01military.html?ref=world.

Dubik, James M. (2009). *Building Security Forces and Ministerial Capacity. Iraq as a Primer.* Washington DC: Institute for the Study of War.

Durch, William J. and Madeleine L. England, Eds. (2010). *Enhancing United Nations Capacity to Support Post-Conflict Policing and Rule of Law.* Washington DC: Stimson Center Report No 63.

Giustozzi, Antonio (2008). "Shadow Ownership and SSR in Afghanistan". In Timothy Donais, Ed. *Local Ownership and Security Sector Reform.* Berlin: Geneva Centre for the Democratic Control of Armed Forces.

Gohel, Sajjan (2012). *Afghanistan: Green-on-blue Attacks Show There's no Easy Way Out.* CNN online 18.9.2012. http://edition.cnn.com/2012/09/18/opinion/opinion-afghanistan-green-on-blue/index.html.

Haug, Jan Erik (2009). *The Operational Mentoring and Liaison Team Program as a Model for Assisting the Development of an Effective Afghan National Army.* Fort Leavenworth, Kans: US Army Command and General Staff College.

Hodes, Cyrus and Mark Sedra (2007). "The Search for Security in Post-Taliban Afghanistan. Chapter Five: Security Sector Reform". *The Adelphi Papers* 47(391), 51–93.

Kelly, Terrence K., Nora Bensahel and Olga Oliker (2011). *Security Force Assistance in Afghanistan: Identifying Lessons for Future Efforts.* RAND Corporation, Arroyo Center. www.rand.org/pubs/monographs/MG1066.html.

Krahmann, Elke (2005). "Regulating Private Military Companies: What Role for the EU?" *Contemporary Security Policy* 26(1), 103–125.

Krekvik, Ola (2011). "Forsvarets samvirke med afghanske styrker. Dilemmaer og utfordringer". [The Norwegian Armed Forces' cooperation with Afghan Security Forces.] *Oslo Files on Defence and Security,* December 2011. Oslo: The Norwegian Institute for Defence Studies.

Lien, Guro 2014. "Militær rådgivning og assistanse." [Military Advising and Assistance.] In Tormod Heier, Anders Kjølberg og Carsten Rønnfeldt, Eds. *Norge i internasjonale operasjoner. Militærmakt mellom idealer og realpolitikk. [Norway in*

International Operations. Military force between idealism and realism.] Chapter 17. Oslo: Universitetsforlaget.

Luck, Edward C. (2006). *UN Security Council. Practice and Promise.* New York: Routledge.

Martens, Kimberly (2012). *Warlords. Strong-Arm Brokers in Weak States.* New York: Cornell University Press.

Miller, Paul D. (2013). *Armed State Building. Confronting State Failure, 1898–2012.* Ithaca, NY: Cornell University Press.

New York Times online (31 March 2004). *Enraged Mob in Falluja Kills 4 American Contractors.* www.nytimes.com/2004/03/31/international/worldspecial/31CND-IRAQ.html?pagewanted=all.

Nilsson, Claes and Kristina Zetterlund (2011). *Arming the Peace. The Sensitive Business of Capacity Building.* Stockholm: Swedish Defence Research Agency, FOI-R–3269–SE.

Norheim-Martinsen, Per Martin, Tore Nyhamar, Anders Kjølberg, Stian Kjeksrud and Jacob Aasland Ravndal (2011). *Fremtidens internasjonale operasjoner.* [The Future of International Operations.] FFI report 2011/01697. Kjeller: Norwegian Defence Research Establishment.

Ramsey, Robert D. III (2006a). *Advising Indigenous Forces: American Advisors in Korea, Vietnam, and El Salvador.* Fort Leavenworth, Kans: Combat Studies Institute Press.

Ramsey, Robert D. III (2006b). *Advice for Advisors: Suggestions and Observations from Lawrence to the Present.* Fort Leavenworth, Kans: Combat Studies Institute Press.

Singer, P. W. (2008). *Corporate Warriors. The Rise of the Privatized Military Industry.* Ithaca, NY: Cornell University Press.

Smith, Rupert (2005). *The Utility of Force. The Art of War in the Modern World.* London: Penguin Books.

Stoker, Donald (2008). *Military Advising and Assistance. From Mercenaries to Privatization, 1815–2007.* New York: Routledge.

Tarp, Kristoffer N. and Frederik F. Rosén (2012). "Coaching and Mentoring for Capacity Development. The Case of South Sudan". *African Security Review* 21(1), 15–30.

US Army Field Manual 3–07.1. (2009). *Security Force Assistance.* Washington DC: US Headquarters, Department of the Army. (Approved for public release).

West, Owen (2012). *The Snake Eaters. An Unlikely Band of Brothers and the Battle for the Soul of Iraq.* New York: Free Press.

Younossi, Obaid, Peter Dahl Thruelsen, Jonathan Vaccaro, Jerry M. Solinger and Brian Grady (2008). *The Long March. Building an Afghan National Army.* RAND Corporation, National Defense Research Institute. www.rand.org/pubs/monographs/MG845.html.

6 Special operations forces – a weapon of choice for future operations?

Iver Johansen

Elite units of the guerrilla, counterguerrilla, and commando type offer politicians in democracies both a tool of policy and a source of fantasy.

(Cohen 1978, p. 101)

Since World War II, forces specifically trained and equipped for unconventional missions have been a feature of the military establishments of most states. Despite being employed in conflict zones worldwide, and at times figuring high in the popular imagination, special operations forces have nevertheless mostly remained a sideshow to the conventional army, navy and air force. The developments since the terrorist attacks on 11 September 2001, however, have radically changed all of that. In the US, but to some degree also in many other countries, special operation force establishments have expanded beyond anything seen before in peacetime or in war. Since 2001, United States Special Operations Command (SOCOM) manpower has nearly doubled, the budget nearly tripled, and overseas deployments have quadrupled (Feickert and Livingstone 2011). In 2014, US SOCOM stood at an all-time high of 63,000 active duty, National Guard and reserve personnel, a number that is planned to continue to increase until the mandated force of 71,000 is reached (Feickert 2014; Thomas and Dougherty 2013, p. 79).[1]

Traditionally, military leaders have been sceptical of the utility of special operations forces, whereas political leaders see in them a tool to solve strategic problems 'on the cheap'. But as Matthew Johnson points out, in the US military and political leaders now agree on the value of such forces for the first time. This development has come about for three main reasons: the 9/11 attacks and the onset of the global war on terrorism; a new security environment marked by irregular threats; and the lessons learned from the employment of special operations forces in Iraq and Afghanistan (Johnson 2006). In addition, successes – notably the killing of Osama bin Laden – have contributed to project special operations forces to the forefront of contemporary warfare.

More generally, the growth in special operations mainly results from two concurrent trends: the continued existence of asymmetric conflict that plays out among the people (Smith 2006), and constraints on defence budgets in the wake of the financial crisis leading to a need to do more with smaller forces. Thus, in asymmetric scenarios typical of the post-cold war environment, small, flexible, highly trained forces able to apply force with great precision offer the promise of achieving strategic results without creating a large and prohibitively costly military footprint on the ground. The defeat of the Taliban regime in Afghanistan in late 2001 was in large measure the result of unorthodox operations by small contingents of US and coalition special operations forces acting in collaboration with local guerrillas. In Iraq in 2003, the rapid conclusion of major combat operations in only 21 days was equally dependent on the largest use of special operations forces since World War II. However, lasting peace has eluded both countries, reflecting the fact that there is no straight line between military effect and strategic effectiveness. The operational excellence of special operations forces thus may not translate into strategic performance, especially if the political goals are ill defined or the means to reach those goals are not available.

In Chapter 2 of this book, Tore Nyhamar identifies five parameters that define future military operations: the importance of justification – usually in the form of a mandate; the complex relationship between consent and partiality; the intensity of conflict; the need to be able to adapt to various operational environments; and relative force composition and strength.

The remainder of this chapter will address these parameters in terms of the development and employment of special operations forces. The point of departure is an analysis of the development of special operations forces – their use, composition and significance – since their origin during World War II and until operations in Afghanistan and Iraq. The focus will be on the unfolding mission portfolio of special operations forces and what that may tell us about their possible roles in future operations. The chapter concludes by discussing where this might lead in the future. Of particular interest is to what extent current trends indicate a deeper transformation of warfare, and whether the capabilities embodied by special operations forces will be required to fight or, indeed, to prevent future wars.

What is special about special operations forces?

What is striking about special operations forces is that they are not specialised in the normal sense of the word, i.e. being specifically trained and equipped to perform one particular set of tasks and to do that at a high level of proficiency and reliability. On the contrary, special operations forces are special because they do things that the ordinary army, navy or air force cannot do, or at least cannot do without incurring high costs and risks of failure. Thus, it is more the *generality* of their abilities than the *specificity* of their tasks that define special operations forces.

Definitions of special operations abound. One oft-quoted definition that still captures the essential aspects of special operations was devised as long ago as the 1980s by Maurice Tugwell and David Charters. According to this definition special operations are first and foremost, 'Small-scale, clandestine, covert or overt operations of an unorthodox and frequently high-risk nature, undertaken to achieve significant political or military objectives in support of foreign policy' (Tugwell and Charters in Barnett *et al.*, Eds. 1984 p. 35). Special operations forces operate in small units. The basic idea at the core of special operations is to achieve large effects with a small input in terms of forces, mainly through extreme precision in the application of force against an opponent's critical vulnerabilities. Furthermore, special operations forces use techniques and operational concepts that are unorthodox, unconventional and not expected by the opponent. Employing force outside the parameters expected enables special operations forces to attain surprise over their opponents. Surprise is the single most important factor that influences battle outcomes, by some accounts equalling a force ratio of up to 2000:1 (Storr 2009, pp. 49–50). Tactically, surprise in special operations supports the achievement of *relative superiority* – the phase in the conduct of an (offensive) operation which is critical to allow a small, lightly armed force to defeat a larger force or a strongly defended target (McRaven 1996).

However, many – if not most – special operations are not of the offensive, direct action kind. Rather, an increasing share of special operations seek to apply force indirectly e.g. through training and assisting a surrogate force or liaising with a foreign government. In the US, since 2012 special operations are officially subdivided into two categories – 'surgical strike' and 'special warfare' (Robertson 2013, p. 11). Doctrinally, the employment of direct means in the form of a surgical strike is not supposed be the main effort in any operation, but rather a means to allow the longer-term special warfare methods to work.

Special operations forces are tools of national policy. Hence, tactical units are usually led from the highest levels of military command and employed to promote specific political or strategic goals. Consequently, special operations forces are regularly used to collect information that may not be acquired through other means, to influence foreign political or military conditions, or – a special task – to rescue hostages. It is, furthermore, an essential feature of special operations that they are as a rule carried out abroad, and whereas such operations in many cases are conducted overtly and with the support of the host nation, they may require infiltrating agents into other countries without the permission of the local government. Such operations therefore have to be either *clandestine* (the existence of the operation is secret) or *covert* (any association between the sponsor and the covert activity can be denied).

Because of their secretive nature, the use of special operations forces is, moreover, in such cases exempt from normal legal procedures for deploying military forces. What these procedures are varies between states. However, few nations publish unclassified documents on these issues. A special case is the US where covert operations lie within the exclusive domain of the Central Intelligence Agency (CIA) and are subject to specific congressional oversight procedures (Kibbe 2004). In other countries, the dividing lines between the special operations and the intelligence communities may be less clear as special operations forces at times engage directly in covert activities in concert with national intelligence services. As a case in point, the so-called E Squadron of the British special operations forces operated covertly in close cooperation with the intelligence service MI6 during the war in Libya in 2011 (Urban 2012).

Special operations are rooted in the worlds of diplomacy and intelligence gathering as much as in the laws regulating warfare. The dependency on internationally sanctioned mandates, therefore, is less than absolute when it comes to deploying special operations forces to a conflict zone. Of course, special units may be deployed as part of a larger force operating on the basis of a mandate authorising the use of military force. However, as often as not, special operations forces are tasked to conduct missions that seek to further purely national political aims. This type of operation may span a very wide spectrum from hostage rescue to various assistance or liaison missions. In such cases, rules of engagement will likely be far more restrictive, oversight more intrusive, and any use of force will in most cases have to depend on the consent and approval of local authorities or partner forces.

Finally, given their strategic role, special operations forces usually have prioritised access to things like cutting edge technology and high quality intelligence. In combination with an especially rigorous selection and training programme such forces are trained and equipped to conduct missions over a wide spectrum of conflict types, intensities, and operational environments. Because of the high range of capabilities in even a small unit and the (relative) independence of other units, special operations forces are more *flexible* than conventional forces. This is one reason why such forces are popular with the political leadership; if the strategy changes, these forces can be redirected immediately.

Use of special operations forces in the past

The post-World War II era

Special operations forces are products of special circumstances. Thus, during World War II it was the strategic predicament of the British army, being restricted to fight the German Wehrmacht only on the fringes of the European theatre of war, that gave rise to the need for commando units to conduct sabotage missions and connect with indigenous partisan forces within occupied countries. By the end of the war, specialised forces of various types had become an institutionalised part of the military establishments of most nations, including the UK and the US (Johnson 2006, p. 275).

After the war, special force establishments in both the US and UK were all but abolished. In Britain, the Special Air Service (SAS), however, lived on after being reborn in 1947. The Royal Marine Commandos also were retained, whereas the Army Commandos were abolished. In the US, most wartime specialised units were abolished, the overarching institutional structure – the Office of Strategic Services – was broken up and the CIA, which took on the responsibility of covert and paramilitary operations as part of a much wider mission portfolio, was established (Johnson 2006, p. 275).

The ensuing cold war with its focus on nuclear deterrence and large-scale war scenarios was not a favorable environment for special operations forces. The 1970s, however, saw the advent of a new threat from international terrorist organisations. In Germany, the consequences of not having a capable counterterrorist capability were tragically exposed at the 1972 summer Olympics in Munich when the Palestinian group 'Black September' took 11 Israeli athletes hostage. The botched rescue attempt, in which all the hostages and one German police officer ended up killed, led directly to the establishment of the German police's counterterrorist unit – 'Grenzschutzgruppe 9' (GSG 9) (Neillands 1998, pp. 209–210). The requirement for a counterterrorist capability led to a renewed interest in special operations forces among governments all over the world. In addition, several highly dramatic and widely publicised hostage rescue operations served to enhance the reputation of specially trained counterterrorist units, notable cases being the Israeli raid on Entebbe in 1976 and the employment the following year in Mogadishu of the German GSG 9 to rescue passengers and crew on Lufthansa Flight 181. In the US, the highly secretive First Special Operational Detachment – Delta, also known as 'Delta Force', was established in 1977, followed three years later by the Navy's equivalent – 'SEAL Team Six'.

Successful rescue operations at Entebbe and Mogadishu demonstrated the value of special operations forces. In the US, however, the turning point in the development of special operations was the 1980 Teheran hostage crisis. In 1979, Iranian activists had taken 53 American hostages at the US Embassy in Teheran. The Carter administration decided on a military rescue mission. An ad-hoc military force was put together comprising elements from all services in addition to civilian agencies. Beyond the complexity of the mission itself, it turned out that the units involved had no previous experience in either planning or execution of joint operations. Operation 'Eagle Claw' ultimately failed; it was aborted when two helicopters broke down and a third crashed with a C-130 transport aircraft at a remote Iranian desert location killing eight US soldiers (Neillands 1998, pp. 215–217).

In the short term, the disaster of Eagle Claw reinforced the opposition to special operations both within the US military and the political leadership. In the longer term, however, the weaknesses revealed by the failed rescue attempt and the need to fix them led to a thorough reorganisation of the entire US special operations establishment. Thus, as early as 1980, the Joint Special Operations Command was established whose mission was to improve coordination between the various special operations forces. However, further chaos in operations like Grenada (1983) made it plain that what was called for was a joint military organisation for all special operations forces 'to ensure adequate funding and policy emphasis for low intensity conflict and special operations' (US SOCOM 2008, p. 6). SOCOM was eventually established on 13 April 1987.

The Gulf War and beyond

The end of the cold war raised hopes of a new peaceful and democratic international order. The stability of the system, however, soon came under threat with an upsurge in ethnic conflicts, terrorism and irregular warfare in the 'conflict crescent' from the Horn of Africa via the Balkans and the Middle East to Central Asia. In most cases these conflict scenarios evolved in the grey zone between war and peace, thus being an almost perfect fit for the capability portfolio of special operations forces.

Still, in the first large-scale military deployment of the new era, Operation Desert Storm, special operations forces were largely left out of the operational plan. On the initiative of the British force commander, General Peter de la Billière, himself a former SAS officer, British SAS troops along with US special operations forces were in the end put onto the strategically important task of hunting down Iraqi Scud missile launchers. After the war, a Pentagon report indicated that the success of these operations might have been more ambiguous than initially claimed (Johnson 2006, p. 282). However, it remains a fact that the Scud strikes against Israel ceased after 24 January, coinciding with the onset of special operations directed against the Iraqi missile capability (Neillands 1998, p. 295).

Desert Storm was not the success the special operations community had looked for, and the deployment of US special operations forces in Somalia two years later turned out to be a direct failure. By 1993, US forces were deployed to support the ongoing UN humanitarian mission. On 3 October, a force consisting of Army Rangers and members of Special Forces Operational Detachment – Delta – launched an operation to capture two leading aides of warlord Mohammad Farah Aideed. The operation ended in disaster when two helicopters were shot down and the rescue teams sent in were pinned down by heavy fire from surrounding buildings. Finally, about ten hours after the start of the operation, a relief column came through, but by that time 19 soldiers were dead or missing and 84 wounded (Adams 1998, pp. 261–264). The Mogadishu operation is the theme for the book and the film *Black Hawk Down*.

Both the Scud hunt and the Mogadishu raid served as examples of the risks and dangers connected with special operations. In the Iraqi desert, the limited results achieved were at least partly due to deficient coordination between special operations and conventional units, and communication between special operations teams and air-support was also lacking (Johnson 2006, p. 282). Mogadishu was more than anything an example of overconfidence in abilities and the misconception that special operations forces 'can accomplish any mission' (Johnson 2006, p. 283). The aftermath of the operation furthermore exhibited the potential for severe political repercussions from a failed mission.

The effects of the Mogadishu raid were still being felt during the 1999 air campaign in Kosovo, when US special operations forces were not allowed to direct air strikes and conduct battle damage assessment, primarily due to force protection concerns. These tasks were instead conducted by French and British forces. The Balkan wars, however, brought developments in special operations in a number of other important respects. As one example, special operations forces from NATO nations engaged widely in the hunt for persons indicted for war crimes. Although often hampered by a climate of mistrust, as the French in particular were accused of harbouring Serb war criminals within their allotted patrolling area in Bosnia, by 1999 15 suspected war criminals had been arrested in Bosnia alone, the majority in operations involving British special operations forces (Zaalberg 2005, p. 292, *Daily Mail* 1999). The operations in the Balkans also provided a useful laboratory for improving integration between conventional and special operations forces at the tactical level (Johnson 2006, pp. 283–284). Thus, many of the lessons learned from a broad mission portfolio, including intelligence gathering, intermingling with the local population and liaising between NATO forces and the various ethnic groups, proved useful for the later employment of special operations forces in Afghanistan and Iraq.

9/11 and the global war on terrorism

The attacks on New York and Washington DC on 11 September 2001 brought fundamental changes to US military strategy and the role of special operations forces within it. Soon after the attacks, it became clear that the Islamist group al-Qaeda was responsible, and that the attacks had been planned and prepared in training camps in Afghanistan. When the Taliban regime in Kabul refused to turn the al-Qaeda leader – Osama bin-Laden – over to American authorities, President Bush ordered its overthrow, the destruction of the al-Qaeda network and the capture or killing of Osama bin-Laden himself.

Over the next few weeks, small teams of CIA operators infiltrated the northern parts of Afghanistan. The basic idea was to employ an 'unconventional warfare' approach to unseat the Taliban from power and to dismantle the al-Qaeda network. According to US doctrine, unconventional warfare entails activities intended to 'enable a resistance movement or insurgency to coerce, disrupt, or overthrow a government or occupying power' (USDOD 2011, Ch. II-9). To do this, the CIA was able to pick up on methods, as well as relationships with key figures, of the so-called Northern Alliance dating back to the struggle against the Soviet occupation during the 1980s. The US military, which had no comparable network to build on, and besides were hampered by internal disagreement on the overall strategy, did not deploy forces to Afghanistan until several weeks later (Schroen 2007). Once in place, the ODA-teams,[2] however, proved essential to direct air attacks against Taliban positions. In a series of engagements in which US air power was used extensively, Northern Alliance forces drove the Taliban out of the major northern city of Mazar-e-Sharif and eventually entered the capital itself – Kabul – on 14 November. Further operations succeeded in reducing Taliban as well as al-Qaeda strongholds in the south also. The entire campaign took fewer than 60 days and involved no more than a few hundred US soldiers and CIA personnel (USSOCOM 2008, p. 101).

As operations in Afghanistan were winding down, the focus soon shifted to the next theatre of major operations – Iraq. In sharp contrast to Desert Storm, Operation Iraqi Freedom involved a substantial special operations component right from the start. Special operations forces operated over virtually the entire Iraqi territory, supporting Kurdish Peshmerga troops in the north, directing air attacks and conducting special reconnaissance on the access routes towards Baghdad (USSOCOM 2008, p. 123). As one example of how special operations forces were used in new and unfamiliar ways, former Delta Force commander Pete Blaber chronicles how a conventional tank unit was attached to a Delta Force unit to simulate a brigade size force and use hit-and-run tactics to create confusion behind enemy lines (Blaber 2008, p. 5).

However, in Iraq as in Afghanistan, unconventional warfare was only a passing moment. As soon as new governments were in place in Kabul and Baghdad, both missions evolved into supporting the regime instead of fighting it. Thus the transition from an unconventional warfare strategy to one of foreign internal defence and counterterrorism meant that special operations forces had to engage the enemy much more directly and much more closely. In terms of US doctrine, both strategies seek to support a host nation through activities that 'protect against subversion, lawlessness, insurgency, terrorism, and other threats to their security, stability, and legitimacy' and to 'render global and regional environments inhospitable to terrorist networks' (USDOD 2011, Ch. II-8–11). The operational focus therefore increasingly turned towards attacking insurgents, especially since the establishment of efficient national security forces remained elusive.

Yet, special operations in Afghanistan and Iraq were widely seen as successful beyond expectations. The ability of trained operators to merge technology with new and innovative tactics seemed to present a model for how future military operations should be conducted. Furthermore, the breakdown of internal political stability that was later to afflict both theatres of war was construed to be the result of an overemphasis on *conventional* military power, not of the efforts of special operations units. The strategic inconclusiveness of both campaigns thus did nothing to weaken the standing of the special operations community.

Iraq and Afghanistan also served as rallying points for other nations that sought to emulate the American special operations model. A number of nations thus contributed special operations teams to serve independently or with US units in both theatres of operation. One example is Task Force K-BAR in Afghanistan, which comprised special operations teams from inter alia the USA, Canada, New Zealand, Norway, Denmark and Germany. As yet another example, in Iraq the securing of the al Faw oil facility in 2003 was achieved by a combined force consisting of US, British and Polish special operations teams (USSOCOM 2008, pp. 108, 123).

Future roles of special operations forces

So what does the future hold? Obviously, there is no simple answer to this question. While it seems a safe bet that special operations forces will maintain a vital role in counterterrorism for the foreseeable future, most other parameters that pertain to the future remain uncertain. Thus, a recent US study states that '...the next chapter in special operations forces history may look as different from the last decade as the post-9/11 era was from the 1990s' (Thomas and Dougherty 2013, p. 46).

In trying to frame possible future outcomes, two factors are paramount: first, the continued existence of a security environment that is dominated by irregular threats, transnational networks and sub-state groups that are able to unbalance entire states, as cases like the 2012–2013 insurgency in Mali demonstrates; and, second, budgetary pressure in the wake of the financial crisis that increasingly forces governments to look for more cost-effective ways to secure national security.

Containing irregular threats

Irregular warfare can usefully be defined as 'a violent struggle among state and non-state actors for legitimacy and influence over a specific population' (Jones 2012, p. 1). Irregular threats typically thrive on the fringes of state-controlled territories and in the ungoverned spaces of the world. Technology, especially advances in access to and use of the Internet, increases the ability of irregular actors to recruit, organise and carry out attacks. In addition, hostile states may themselves generate irregular threats by sponsoring terrorist groups and facilitating attacks on other countries. Fuelled by social unrest and weak governance, irregular warfare will probably remain a feature of international politics for the foreseeable future. It will also most likely have a large impact on the future role of special operations forces in Western security strategy.

Irregular conflict is not played out purely at the low-intensity end of the conflict spectrum. In Chapter 2 of this book, Tore Nyhamar states that militarily inferior actors will typically seek to compensate for their weaknesses by developing tactical situations that enable them to dominate the battlefield. Therefore irregular actors might employ any means and methods within their grasp, including guerrilla warfare, terrorism, subversion, and insurgency. Thus, what on a strategic or operational level may be considered low intensity, on the tactical level might be experienced as very high-intensity fighting indeed.

Conventional forces are notoriously inefficient at solving the dilemma between acquiescing and unleashing the full force of their military means. The ability of irregular actors to blend in with the local population adds to the difficulties of conventional forces in fighting irregular wars. The answer, in many cases, is to become irregular oneself. One way of becoming irregular is to stop killing enemies and start applying a more indirect approach to the fight, something that goes to the very essence of the special operations forces' operational repertoire. Indeed, James Kiras asserts that special operations are the application of irregular warfare concepts and principles by conventional military powers (Kiras 2006, p. 69).

Irregular warfare, however, spans a wide spectrum of strategies. In confronting irregular actors, an intervening force will have to consider two main parameters – consent and the intensity of conflict as this affects different modes of fighting.

Consent may be given by one or both of the conflicting actors, the international community, or there may even be no consent given at all. For practical purposes, however, we will consider consent on the part of either the local regime or the local population. Modes of fighting can be direct, as when force is applied actively against an opponent, or indirect as when the main effort is directed towards assisting and supporting a – regular or irregular – partner force. Within this framework, four main strategies are possible.

Consent regime/direct mode: This alternative broadly represents US counterinsurgency strategy as it has been applied in places like Iraq and Afghanistan. Counterinsurgency represents a population-centric approach (Jones 2012, p. 6). The intervening force may apply a two-pronged method by shielding the population and attacking the insurgents directly at the same time. The basic idea is to deprive the insurgents of their hold on the local population. The assumption is that the population will opt for the government if the threat from the insurgents is sufficiently reduced. Counterinsurgency is not strictly a special operations task, especially as it presupposes the deployment of very large numbers of outside forces. However, although special operations forces may have a role to play within a counterinsurgency campaign, considering the dismal results of the vast counterinsurgency efforts in Iraq and Afghanistan this particular approach does not seem to be the most likely path forward in future special operations.

Consent regime/indirect mode: This alternative focuses primarily on advising and assisting the host nation's security forces and enabling those forces to constitute the front line in the fight against the insurgent forces. This strategy falls within the definition of foreign internal defence in US doctrine, or military assistance in NATO doctrine (USDOD 2011; NATO 2009). Foreign internal defence has been a feature of US strategy in both Afghanistan and Iraq, although the direct mode of fighting tended to trump the employment of indirect means in both theatres of war. Other examples of a more purely indirect nature, however, include Operation Enduring Freedom – Philippines, where US forces played a supportive role only, leaving the actual fighting to the Philippine armed forces themselves (Maxwell 2013). An in-depth analysis of military advising and assistance operations can be found in Chapter 5 of this book. Playing more directly on the indirect approach by engaging local partner forces, this approach may avoid some of the pitfalls connected to a more direct application of military force. Thus, in the future, special operations may tend to be more indirect, more of the assistance type and hence more long-term.

Consent population/direct mode: This alternative implies actively taking part, as well as supporting an insurgent force in the fight against a hostile government. This strategy corresponds to unconventional warfare in US military terminology. It builds on the active use of armed force to destroy the opponent and gain control over vital territory, including the capital and the state institutions. The US campaign against the Taliban in 2001 represented a classic example case of an unconventional warfare campaign. Another case was the war in Libya in 2011, where NATO special operations forces were employed covertly to increase the effectiveness of air delivered munitions, to train officers and, possibly, also to supply weapons to anti-regime forces (Urban 2012). In cases where large-scale deployment of regular forces is not an option – and it is very likely that in the majority of future cases it will not be – special operations forces will increasingly be called upon to perform key tasks to coordinate various military efforts by outside actors, including the use of air power.

Consent population/indirect mode: In a classic guerrilla war scenario, an intervening force may opt for a more indirect approach to support an insurgency in its fight against qualitatively superior government forces. This strategy is not directed towards quick results but accepts that the strategy will have to be applied over the long term. As such, the strategy rests on the principle of attrition; the purpose is not to disarm the enemy but rather to erode his will to continue the fight. In Maoist theory, revolutionary war consists of three stages: the strategic defensive, the stalemate, and the strategic offensive (Kiras 2006, p. 71). Only when the opponent's forces are worn down and his morale weakened is it time for the final and decisive offensive. Unconventional warfare in the indirect mode has been applied in such diverse scenarios as occupied Europe during World War II and Afghanistan under Soviet occupation until 1989. Although guerrilla warfare will remain within the special operations forces' portfolio, in the absence of a 'peer competitor' on the global scene, it is not very likely that this particular approach will figure prominently among Western special operations forces in the future.

The findings are summarised in Table 6.1 below.

It is important to note that these strategies are ideal types, and that pure models may not be applicable in real world situations. Yet, although methodological flexibility might be required, that is no guarantee against building strategic decision-making on weak or erroneous assumptions. Two possible pitfalls seem to be particularly significant. First, military operations to stabilise a country or a region where an insurgency is ongoing may fail if there is little in terms of host nation structures and support for an intervening force to build on. History carries overwhelming evidence that it is a waste of effort to support a corrupt regime that does not have the trust and confidence of the population. As Seth Jones puts it, 'an outside power cannot force a local government to be legitimate' (Jones 2012, p. 6).

Second, even if the basic conditions for success are in place, the dynamics of conflict and the desire for quick results may draw attention and resources towards direct approaches for short-term results at the expense of potentially more effective long-term indirect uses of military force. The essential problem is that the distinction between direct and indirect modes of fighting tends to be clearer in theory than it is in practice. In terms of special operations, the basic idea is that the direct approach 'buys time' (Robertson 2013, p. 13) for the indirect approach to work decisively. For instance, it may be necessary to actively pursue and neutralise a threat to enable political–military activities by special operations forces (and perhaps other agents of national policy) to take place that may address the underlying problems in a more lasting manner. In this overall design, the indirect activities are supposed to be decisive, whereas the direct activities are supposed to be supportive and ad hoc.

In practice, though, short-term operational requirements have tended to trump more long-term uses of special operations forces, redirecting resources increasingly towards the direct-mode end of the operational spectrum. James Kiras maintains that when special operations forces have been used tactically against numerically superior foes, 'the response historically has been to expand organizations at the expense of overall quality' (Kiras 2006, p. 67).

Table 6.1 Irregular warfare strategies

		Consent	
		Consent regime	*Consent population*
Mode of Fighting	Direct Mode	• Counter-insurgency • Afghanistan/Iraq • Population centric	• Unconventional Warfare • Afghanistan 2001/Libya • Support insurgent force
	Indirect Mode	• Military Assistance/Foreign Internal Defence • OED – Philippines • Assist government forces	• Guerrilla tactics (Maoist) • Occupied Europe WW II • Afghanistan pre-1989 • Long-term attrition

As a case in point, over time special operations in Afghanistan have been raised to industrial levels. According to Robert Fry, who was deputy commanding general for coalition forces in Iraq in 2006, special operations forces are now supported by '... an army of military support groups, linguists, analysts, interrogators, computer and forensic geeks, all linked from the operating theatre by huge data pipes to the parent intelligence agencies in the US and UK" (Fry 2012, p. 32). This 'economy of scale' model enables a vast increase in the production of special missions, and though it might have been necessary to produce the one that took down Osama bin Laden, it may not be a formula for the optimum employment of special operations forces to produce long-term results.

What is needed is a model for how to use special operations forces to best effect – to do what no other military force can achieve (Robertson 2013, p. 14). The crucial issue is to avoid spending scarce resources in a tactical and ad hoc manner, rather than as one among a number of tools in a comprehensive effort to achieve decisive and lasting outcomes.

An affordable solution to strategic problems

As noted in the introductory chapter, low overall economic growth in most Western countries after the financial crisis has put increasing pressure on defence budgets. In Europe, most middle-sized countries have implemented defence spending cuts of 10 to 15 per cent on average. Leading countries like the UK, Germany and France are reducing personnel strength as well as cutting budgets, in the case of the UK by up to 25 per cent in real terms over a four year period (O'Donnell 2012, Erlanger 2013). The US defence spending is down by about 10–15 per cent from a high in 2010, mainly due to Iraq war reduction and a scale-down in Afghanistan (Shah 2013). In January 2012, the Obama administration issued new defence strategic guidance that, among other things, demanded the adoption of 'innovative, low-cost, and small-footprint approaches to achieve security objectives' (quoted in Robertson 2013, p. 5).

The strategy of countering terrorism and other evolving security threats through large-scale force deployments is no longer economically viable. Going to smaller forces, however, may save money but risks creating an even more disadvantageous ratio between the size of the intervening force and the population. In Chapter 2, a force ratio of about 20 soldiers per 1,000 population is indicated as a necessary minimum to fight and prevail against an insurgency. Even counting indigenous forces, that kind of force ratio has mostly not been met in major conflict scenarios after World War II, including Iraq and Afghanistan.

At the same time, a lowering of operational objectives may not be an option in the future. Given the widening gap between operational requirements and available resources, shifting the burden of fighting irregular wars to special operations forces may sound like a panacea. Special opera-tions forces represent high costs per soldier but low cost overall. And if results compare with, or even exceed, what may be accomplished by larger conventional force deployments, that may look like exactly the kind of cost-effective solution governments are seeking. Thus, in an attempt to do more with less, the new model that is now being advocated by Admiral W. H. McRaven, commander US SOCOM, emphasises 'persistent pres-ence' by special operations forces in exposed regions to 'gain expanded situational awareness of emerging threats and opportunities' (quoted in Thomas and Dougherty 2013, p. 81). The new model also implies a reshaping of America's special operations forces towards the indirect end of the operational spectrum. In the future, special operations forces will 'increasingly conduct operations short of war that are more indirect and less kinetic to confront a variety of interconnected, cross-border chal-lenges to include: localizing and defeating VEN (Violent Extremist Net-works) across a number of continents, waging long-duration influence campaigns and proxy competitions in multiple regions and key states, and interdicting WMD (Weapons of Mass Destruction'. (Thomas and Dough-erty 2013, p. 78).

The ability to identify and target emerging threats is dependent on vast partner networks. Thus, the building of a global special operations forces network consisting of foreign allies as well as interagency partners stands as the centrepiece of the model. The training of partner forces will be the task of the US Army Special Forces, so that for the next crisis a capable local force will be available right from the outset. Building capable local forces might also reduce the dependence on counterterrorist forces like Delta Force or SEAL Team Six to engage terrorists and other threats directly.

The new plan, however, comes with a number of question marks. On the one hand, shifting the burden of containing new threats from general-purpose forces to special operations forces will increasingly strain even the vast US special operations establishment. Already the demand for special operations forces in the US outstrips available supply. According to General J. F. Mulholland, commander of US Army Special Operations Command 2008–2012, the Command 'can only satisfy about 50 percent of demand out there … for special operations forces' (quoted in Thomas and Dougherty 2013, p. 44). Any policy that seeks to expand and diversify special operations tasks and missions will thus have to contend with the 'truth' that special operations forces cannot be mass produced.

Also, increasing the role of SOCOM will likely provoke opposition from other parts of the US military as well as from governmental agencies. Designed to uphold US engagement globally in an era with shrinking defence budgets, the consequence in all likelihood will be less money for the maintenance of general-purpose forces while SOCOM is allowed to continue to grow. Alternatively, SOCOM will have to find the money for developing its global network from within its own budget. In fact, SOCOM has already been ordered to realise the global network in a resource neutral fashion (Feickert 2013), indicating that there will be tighter limits to the influx of money into the US special operations forces' budgets in the future.

Moreover, law-makers grow more and more wary of the potential weakening of political oversight and accountability that comes with turning the responsibility for conducting additional operations, including covert operations, over to the military (McLeary 2013). Lacking a legislative framework, the ability to engage partner forces in operations that may have to be conducted covertly or clandestinely may thus be severely restricted. In addition, even though Admiral McRaven has insisted that his network will not act without approval from local US ambassadors, this might not be enough to reassure the foreign service or, indeed, foreign governments (Ignatius 2013).

Whereas the global network approach might be a feasible strategy for US special operations forces, this will probably be far less important for special operations forces of other nations. Lacking the vast resources – and the global ambitions – of the US special operations establishment, other nations' forces will have to continue to do what they have always done. This means maintaining capable counterterrorist forces of the SAS or GSG 9 type while at the same time using those forces for other tasks when that is needed. It is probably only the US that can afford to maintain substantial forces for direct action tasks while at the same time building networks of allies and partners globally.

The overall effectiveness of the network concept, furthermore, rests on the ability to tap into local institutions or to be able to provide the capacities and knowledge that are needed to build those institutions. However, entrusting local partners with training and materiel might be a double-edged sword. A case in point might be found in the deployment of special operations forces in Mali ahead of and during the 2012–2013 uprising. Although the US did not deploy troops to Mali – that was being done by France and the UK with US logistical support – the Mali army had for a number of years been the recipient of US military aid and training (Axe 2013). It thus serves as a reminder of the risks connected with this kind of preventive engagement that it was the same troops the US had been training who participated in the overthrow of the elected government. Furthermore, the US training programme had in part relied on ethnic Tuareg to command Malian elite units and they, in turn, defected to the insurgency in the north, robbing the collapsing government troops of leadership, weapons and equipment (Baumann 2013). According to US Africa Command (AFRICOM) commander General Carter Ham, what went wrong in Mali was 'that US training focused almost exclusively on tactical and technical competence and perhaps not enough on values, ethics and military ethos' (Warner 2013, p. 6).

Conclusion

Ever since the onset of the global war on terrorism in the aftermath of 9/11, pundits have been heralding a new 'era of the operator' (Dorschner 2013). The strategic reach, combined with the independent direct action capability and the small footprint of special operations forces contrast starkly with the costly, burdensome and essentially unsuccessful deployments of large conventional forces. Thus, in scenarios like Iraq and Afghanistan, they have proved their worth in hunting down terrorists and combatting guerrilla fighters. However, excellence at the tactical and operational levels is not easily translated into strategic effect. The future relevance of special operations forces therefore depends on the ability to redefine the link between tactical employments and longer-term strategic aims.

In the future, special operations forces will have to contend with two major challenges: to combat terrorists and to fight irregular wars. On both counts, the special operations strategy being developed by the US builds on an increased employment of special operations forces both directly and indirectly by engaging, building and assisting partner forces in exposed regions. Although the establishment of the global network of special operations forces may run into greater difficulties both financially and legally than perhaps expected by its sponsors within the US military, the outcome will most likely be a greater focus on indirect approaches, assisting and liaising and long term deployments. The future will thus see less of the commando-style direct action mode of employment, which may be spectacular and create short term results but which does not necessarily create the strategic result that is sought, and more of the military assistance, civil affairs, advisory and diplomatic efforts, which may stand a stronger chance of providing long-term strategic results.

Special operations forces will also be given key roles as Western forces increasingly fight future wars from a distance. Gathering of information, designation of targets and coordination of other forces can probably only be done by inserting operators on the ground, and there is no alternative other than to employ special operators for these tasks. Thus, while special operations forces have not redefined warfare, nor will they replace or make superfluous nations' general-purpose forces, they will remain indispensable as a force to solve tasks that no other force can handle.

Notes

1 The figure includes Department of Defense (DOD) civilians assigned to US SOCOM Headquarters and its subordinate components.
2 Operational Detachment – Alpha (ODA) is the basic unit of the US Army Special Forces.

References

Adams, T. K. (1998). *US Special Forces in Action. The Challenge of Unconventional Warfare.* London: Frank Cass.

Axe, D. (2013). "British and French Commandos Take Charge of Mali War". *Wired.* www.wired.com/dangerroom/2013/01/mali-commandos/.

Baumann, A. (2013). "Shifting Parameters of Military Crisis Management". *International and Security Network. Swiss Federal Institute for Technology Zürich.* www.isn. ethz.ch/isn/Digital-Library/Articles/Special-Feature/Detail/?lng=en&id=16129 0&contextid774=161290&contextid775=161268&tabid=1454208072.

Blaber, P. (2008). *The Mission, the Men and Me. Lessons From a Former Delta Force Commander.* New York: Berkley Caliber.

Cohen, E. A. (1978). *Commandos and Politicians. Elite Military Units in Modern Democracies.* Cambridge, Mass: Center for International Affairs, Harvard University.

Daily Mail (1999). "SAS Swoop on Serb Butcher'. *Daily Mail* 21 December. www.sas-specialairservice.com/sas-kosovo-serbia.html.

Dorschner, J. (2013). "The era of the operator". *Jane's Defence Weekly*, 27 March.

Erlanger, S. (2013). "Grim Economics Shape France's Military Spending". *New York Times.* 30 April. www.nytimes.com/2013/04/30/world/europe/grim-economics-shape-frances-military-spending.html.

Feickert, A. and Livingstone, T. K. (2011). *US Special Operations Forces (SOF): Background and Issues for Congress.* www.fas.org/sgp/crs/natsec/RS21048.pdf.

Feickert, A. (2014). *US Special Operations Forces (SOF): Background and Issues for Congress.* Washington DC: Congressional Research Service, May 8, 2014.

Fry, R. (2012). "Survival of the fittest". *Prospect Magazine.* www.prospectmagazine. co.uk/magazine/special-forces-sas-future/.

Ignatius, D. (2013). "Drawing Down, But Still Projecting Power". *Washington Post.* www.washingtonpost.com/opinions/david-ignatius-drawing-down-but-still-projecting-power/2013/03/29/591ebe30–9895–11e2–814b-063623d80a60_story. html.

Johnson, M. (2006). "The Growing Relevance of Special Operations Forces in US Military Strategy". *Comparative Strategy* 25(4), 273–296.

Jones, S. G. (2012). *The Future of Irregular Warfare.* www.rand.org/content/dam/ rand/pubs/testimonies/2012/RAND_CT374.pdf.

Kibbe, J. D. (2004). "The Rise of the Shadow Warriors". *Foreign Affairs.* March/ April, 102–115.

Kiras, J. D. (2006). *Special Operations and Strategy. From World War II to the War on Terrorism.* London: Routledge.

Maxwell, D. S. (2013). "Partnership, Respect Guide US Military Role in Philippines". *World Politics Review.* www.worldpoliticsreview.com/articles/12685/ partnership-respect-guide-u-s-military-role-in-philippines.

McLeary, P. (2013). "Lawmakers Skeptical of Global Spec Ops Plan". *Defense News.* www.defensenews.com/article/20130810/DEFREG02/308100007/Lawmakers-Skeptical-Global-Spec-Ops-Plan.

McRaven, W. H. (1996). *Spec Ops. Case Studies in Special Operations Warfare: Theory and Practice.* New York: Ballantine Books.

NATO (2009). Allied Joint Doctrine for Special Operations.

Neillands, R. (1998). *In the Combat Zone. Special Forces since 1945.* New York: New York University Press.

O'Donnell, C. M. (2012). *The Implications of Military Spending Cuts for NATO's Largest Members.* Washington DC. www.brookings.edu/~/media/research/files/papers/2012/7/military spending nato odonnell/military spending nato odonnell pdf.

Robertson, L. (2013). "The Future of US Special Operations Forces" in *Council on Foreign Relations Special Report No. 66.* Washington, DC: Council on Foreign Relations.

Schroen, G. C. (2007). *First In. An Insiders Account of How the CIA Spearheaded the War on Terror in Afghanistan.* New York: Ballantine Books.

Shah, A. (2013). "World Military Spending". *Global Issues.* www.globalissues.org/print/article/75#USMilitarySpending [Accessed 4 December 2013].

Smith, R. (2006). *The Utility of Force. The Art of War in the Modern World.* London: Penguin Books.

Storr, J. (2009). *The Human Face of War.* London: Continuum.

Thomas, J. and Dougherty, C. (2013). *Beyond the Ramparts: The Future of US Special Operations Forces.* www.csbaonline.org/publications/2013/05/beyond-the-ramparts-the-future-of-u-s-special-operations-forces/.

Tugwell, M. and Charters, D. (1984). "Special Operations and the Threats to United States Interests in the 1980s". In F. Barnett, H. Tovar and R. Schultz, Eds. *Special Operations in US Strategy.* Washington DC: National Defense University Press, 27–44.

Urban, M. (2012). "Inside Story of the UK's Secret Mission to beat Gaddafi". *BBC News Magazine.* www.bbc.co.uk/news/magazine-16573516.

USDOD 2011. Joint Publication 3–05 Special Operations. Washington DC: Department of Defense.

USSOCOM (2008). *United States Special Operations Command History.* www.fas.org/irp/agency/dod/socom/.

Warner, L. A. (2013). "Capacity-Building Key to Africom's Mission". *World Politics Review.* www.worldpoliticsreview.com/articles/12689/capacity-building-key-to-africoms-mission.

Zaalberg, T. W. B. (2005). *Soldiers and Civil Power: Supporting or Substituting Civil Authorities in Peace Operation during the 1990s.* University of Amsterdam. http://dare.uva.nl/document/102314.

7 The future of UN peacekeeping operations

Stian Kjeksrud

Most military UN peacekeepers currently operate in hostile conflict environments in Africa. In Mali, the DRC, Abyei, South Sudan, Darfur, Liberia, the Ivory Coast and the Central African Republic, Blue Helmets are tasked with implementing highly ambitious mandates. How these troops operate on the ground is shaped by an intricate interplay between many actors. The five permanent members of the UNSC, the UN Secretariat, UN mission headquarters, the troop-contributing countries, host authorities, non-governmental organisations, national security forces, armed groups, civilians, the Blue Helmets themselves and many others constantly alter the organisation's approach to peacekeeping operations. Adding to this complexity is the sheer scale of UN deployments, as close to 100,000 uniformed personnel, in collaboration with almost 17,000 civilian personnel, are deployed to 16 operations with widely different mandates on four continents (United Nations 2014a). Any attempt to capture the essence of modern UN peacekeeping operations will therefore struggle to provide clarity, and efforts to predict their future trajectory will at best be imprecise.

Despite the obvious pitfalls encountered when presenting an imagined future, this chapter will provide a fore*sight* (see Chapter 1) into the near future of UN peacekeeping operations. The main point is to portray a selection of possible developments, observable in various degrees today, that may impact the conduct of future UN operations. It will do so through the lens of the parameters presented in Chapter 2. Ideally, the discussions will inspire current and future troop contributors to improve their ability to manage the uncertainty which comes with a changing global environment. The chapter will thus describe a not so distant future where the UNSC moves towards more limited and more robust UN peacekeeping mandates (*mandate*). In these operations, UN troops will take sides and, at times, operate without the consent of the main parties of conflict while faced with violent opposition to the UN's use of force (*consent* and *conflict intensity*). This will unfold in some of the most demanding operational environments in the world. Asian and African infantry forces, supported by small numbers of European peacekeepers equipped with niche capabilities, will be deployed in parallel or even jointly with regional organisations and influential countries (*operational environment* and *relative force composition and strength*). It concludes by pointing to a need to revise the bedrock principles of UN peacekeeping operations and suggests how troop contributors may prepare for future peacekeeping deployments.

Rupert Smith has argued that a consequence of modern warfare – where wars are fought about, and amongst the people – is that we must rethink how military force can be utilised and organised to better support new political objectives, i.e. improve the utility of force (Smith 2006). He also states that the UN, due to inherent military and political constraints at the strategic level, will 'never offer a serious option for the use of military force' (Smith 2006, p. 18). Maybe for this reason, Smith's paradigm on modern war is rarely used in studies of military challenges in UN operations. However, this chapter claims that the UN has much to learn from Smith. One simple reason for this is that the organisation deploys close to 100,000 uniformed personnel worldwide to address some of the most difficult and bloody 'modern' conflicts, relying heavily on military forces in order to succeed. Given the organisation's complicated relationship with the *use* of armed force, Smith's work may help us rethink how military force may be used with more utility in future peacekeeping operations. Accordingly, some of his insights will be used throughout the chapter to highlight some issues in particular.

Towards more limited and robust UN mandates

A move towards more limited and robust mandates for UN peacekeeping operations will come as a reaction to the highly ambitious nation- and peace-building agendas dominating many of the organisation's mandates for the last ten to 15 years. These ambitions, having led to a massive expansion in both mandated tasks and UN troops on the ground, put a strain on the organisation's ability to plan, deploy and sustain peacekeeping operations. They also made 'success' much harder to achieve. The more limited mandates for UN operations in Abyei (contested area in Sudan/South Sudan) in June 2011 and Syria in April 2012 may indicate that the UN had then reached a point at which ambitions were being scaled down. From one point of view this may be seen as pragmatic solutions to remedy symptoms of overstretch, from another as a consequence of friction within the Security Council. UN engagements in Mali (United Nations 2013a), the Eastern DRC (United Nations 2013b), and the Central African Republic (United Nations 2014a), however, paradoxically seem to indicate a parallel development where the use of military force to protect civilians, deter armed groups and stabilise population centres will also dominate future UN missions.

With the end of the cold war, the UN was intended to have a greater role in promoting democratic change in failed states but was soon met with colossal challenges. It became clear that UN peacekeeping operations were not calibrated to tackle armed resistance. UN peacekeeping has been in constant reform mode ever since the failures to find proper military responses to mass violence in Rwanda, Bosnia and Somalia. Since then a lot of introspection and reform effort have gone in to making peacekeeping a more effective tool, from the strategic level in New York down to the tactical level in-mission (Carlsson *et al.* 1999; United Nations 1999, 2000, 2001, 2004, 2005, 2006a, 2006b, 2006c, 2006d, 2006e, 2007a, 2007b, 2008a, 2008b, 2008c, 2009a, 2009b, 2009c, 2010a, 2010b, 2010c, 2010d, 2010e, 2011a, 2011b). A landmark effort was the so-called Capstone Doctrine, published in 2008, where, for the first time, 60 years of peacekeeping experience was synthesised to provide specific guidance to modern day peacekeepers (United Nations 2008c). In mid-2010, the then head of the UN Department of Peacekeeping Operations (DPKO), Alain le Roy, stated that the UN had reached a 'consolidation phase'. Reform efforts were now to be focused on *refining* existing concepts and approaches (UN News Centre 2010).

Not only has the UN strived to do what it does better, but it has also tried to do a lot more. What may be portrayed as *the* defining development of UNSC mandates over the last decade is the expansion of ambition, best portrayed by the wide array of tasks given to both military and civilian peacekeepers. The Security Council now typically authorises UN operations to address drivers of conflict across *all* sectors of society by reforming, rebuilding or even creating core mechanisms composing the inner workings of the host state. Many of the tasks found in these ambitious mandates lie well outside what military troops traditionally have been prepared to perform (Heinecken and Ferreira 2012). Following these changes to UN mandates and operations, the organisation has made steps towards better integration of UN efforts. The UN now deploys *integrated missions* in attempts to avoid internal UN overlap, unnecessary friction with host states and to provide more effect on the ground (Eide *et al.* 2005).

Despite well-intended reform processes and new, sometimes innovative, ways of organising operations, positive impact on the ground has been harder to discern. Thus, the question of whether the UN is overstretched – politically, strategically and militarily – has emerged. This issue has been addressed from several angles in many studies and reports over the past few years and many have argued that the UN has indeed reached the point at which ambitions clearly outweigh the organisation's actual ability to follow through (Tardy 2007, 2011; Gowan 2008, 2009, 2012a; Kjeksrud 2009).

The 'consolidation phase' referred to by le Roy in 2010 is possibly heading towards a period of contraction. Richard Gowan provides three reasons why 'the grand narrative of state-building is now breaking down'. Firstly, experience from Afghanistan and DRC questions whether 'states can be built at all'. Secondly, the financial dip constrains 'all organisations' ability to sustain large-scale operations'. Finally, growing political differences between and within the major organisations 'may place limits on what large-scale missions will be able to achieve in the future' (Gowan 2012b). Therefore, the UN will have to rethink what its operations can and should do short of attempting to rebuild states. Some of the most recent UN peacekeeping mandates, especially UNSMIS and the United Nations Interim Security Force for Abyei in South Sudan/Sudan (UNISFA), seem to signal change in Security Council responses to threats to international peace and security (United Nations 2012a, 2011d).

The conflicts in Abyei include both inter-communal violence between the nomadic Misserya and the agro-pastoralist Ngok Dinka and the inter-state conflict between Sudan and newly independent South Sudan over the future status of Abyei. The inter-communal conflict between the Misserya and Ngok Dinka is exacerbated by the fact that both Sudan and South Sudan use the two parties as proxies to influence the unresolved issue of the future status of Abyei. The immediate trigger for UNISFA's deployment to Abyei in June of 2011 was a Sudanese military offensive in the spring of 2011, which drove over 100,000 Ngok Dinka out of the disputed region. Since then, continuous international efforts have been made to reconcile Sudan and South Sudan on the final status of Abyei. Abyei's population is still awaiting the opportunity to hold a referendum to decide whether they should secede to become part of South Sudan or stay connected to Sudan. The fate of the Abyeians is one of the most 'combustible' of the unfulfilled provisions of the Comprehensive Peace Agreement from 2005 (Center on International Cooperation 2012).

UNISFA's core tasks are reminiscent of a bygone era of peacekeeping. In short, its core tasks are to serve as a buffer between warring parties to avoid a slide back to war and to monitor the demilitarisation of Abyei. However, there are contextual aspects that make UNISFA less similar to 'old-school' peacekeeping. First, the mission is deployed to a highly contested area. There are still strong incentives to use violence to resolve the remaining political issues. In addition, UNISFA's mandate includes provisions to protect civilians. This implies that if civilians are threatened in areas where the mission is deployed, UNISFA is supposed to take action, including through the use of armed force. Any armed action by UN forces to protect civilian lives will challenge their impartiality in the eyes of the warring parties. This is particularly sensitive in communal conflicts, since both sides usually alternate between the roles of victims and perpetrators, depending on the point reached in the cycle of communal violence (Beadle 2014).

In this era of *multinational* operations, UNISFA also stands out by being *mono-national*, as the operation is mainly run and staffed by one country only – Ethiopia. Ethiopia provides almost all uniformed personnel and military units, approximately 4,000 troops in total. This set-up may be explained by the fact that Ethiopia had the political will and the available capabilities at the right time. It could also be interpreted as a response which deliberately avoids deploying a complex multinational UN operation. Another novel aspect of UNISFA in this era of *multidimensional* operations, where civil–military–political integrated missions have dominated operational theatres, is that UNISFA is largely a *one-dimensional* military operation. Since Abyei is not a state to be rebuilt, it is quite logical that UNISFA does not need the wide array of civilian components found in more complex UN missions elsewhere. But it also shows that the UN is now willing to deploy a dominantly *military* peacekeeping force to stabilise an area and protect civilians. Compared with the massive expansion in tasks given to UN operations over the last decade, the mandate of UNISFA is thus quite limited. However, more limited does not imply less challenging. As of early 2014, there is no solution in sight to the final status of Abyei, and there is only marginal progress on the implementation of the technicalities that are meant to serve the temporary management of the border areas between the two states.

Another peacekeeping operation with more *limited* ambitions was UNSMIS. At first glance, UNSMIS also invokes memories of former eras of peacekeeping, this time involving monitoring, observing and reporting on a ceasefire agreement. Again, context does matter. In the midst of the violently escalating conflict in Syria, it soon became clear that UNSMIS was operating without the de facto consent of the warring parties. The unarmed military observers quickly became targets for all sides of the conflict and right from the outset were actively hindered from completing their tasks. After several violent incidents which threatened the lives of the observers, UNSMIS activities were suspended just two months into its deployment. On 19 August 2012, the Security Council decided not to renew its mandate for UNSMIS.

There are many aspects that may explain why UNSMIS became an ineffective tool. First of all, peacekeeping's chance of success fundamentally rests on the political will of the permanent five members of the Security Council (P-5). Despite the reform processes over the last 15 years, a peacekeeping mandate will only be as strong as these actors want it to be. A troubling aspect with UNSMIS is that it seems that the 'never again' mantra, which has been able to drive UN reform efforts forward since the failures to respond to the genocide in Rwanda and ethnic cleansing in Bosnia, did not seem to move the Security Council to a more *relevant* response in this instance. What clearly was and remains a situation where UN members states' responsibility to protect the civilian population in Syria applies does not manage to outweigh the political differences in the Council. The Syrian conflict reminds us of the political realities which govern the interplay between the P-5, all wielding veto-power. Disagreements in the Council will be able to outweigh efforts that have been made to make peacekeeping a more effective tool in the future also. This implies that future peacekeepers risk being deployed to conflicts they have little chance of influencing and yet again become passive bystanders to human rights violations against civilians. These are the kind of incidents which have haunted the organisation for the last 20 years.

If the UN steps back from the ambitious peace- and state-building agendas and moves towards deployments where military peacekeepers are set to perform more 'traditional' military tasks, the complexity of the civilian–political–military interplay found in current integrated UN missions will possibly be reduced. In theory, this would allow military forces to do what they do best and may increase the chances of success (although within more limited ambitions). However, as Rupert Smith's insights indicate, modern wars will be fought among the people. Therefore UN forces will probably continue to be deployed to stabilise intra-state conflicts which cannot be resolved through military force alone, if at all. Thus, many of the current challenges facing military peacekeepers will remain. Future UN peacekeepers will have to relate to multifaceted political processes and cooperate with a wide variety of civilian and military international and local actors.

Further, more robust mandates could lead to more expansive use of force by UN peacekeepers, which most certainly will lead to unintended consequences. In Rwanda, Bosnia and Somalia, UN forces were often criticised for not applying enough force, or not the right function of force, to protect civilians. In order to avoid repeating past failures, the UN is now seemingly prepared to act more robustly. Mandates for the Force Intervention Brigade (FIB) in DRC, United Nations Multidimensional Integrated Stabilization Mission in Mali (MINUSMA) and United Nations Multidimensional Stabilization Mission in the Central African Republic (MINUSCA) all point in this direction. However, applying too much force may put UN peacekeeping out of business altogether, as legitimacy may be lost and consent can quickly be withdrawn. A move towards limited and robust mandates will therefore demand a UN that is able to utilise military force with more utility than is the case today.

Whither the bedrock principles of UN peacekeeping operations

If future peacekeepers are expected to wield the tool of military force with better effect, the UN needs not only to rethink its relation to the use of force but also to rethink what that may mean for the bedrock principles of UN peacekeeping: consent, impartiality and the minimum use of force. They have proved essential in former eras of peacekeeping, most of all to underline the UN's rather limited military role in international conflicts, but what is often forgotten is that these principles were developed as a pragmatic necessity during the political climate of the cold war. Today, however, due to the ways wars and conflicts unfold and the increasingly 'deep' interventions by the UN, they are often stretched beyond recognition. They will probably not survive in their current form if the UN moves towards more limited and robust operations.

The principle of *consent* implies that the UN will not deploy an operation without the explicit request of a host nation and the main parties to the conflict. This formal acceptance by the former belligerent parties, usually secured during peace negotiations, is supposed to ensure that the UN operation can perform its mandated tasks without negative interference and with a presumed willingness to support a transition towards peace and stability. Consent often fluctuates, however, and UN operations often find themselves managing consent throughout their lifespan (United Nations 2008c, p. 32). Also, it is often unclear who the main parties to the conflict are, as violent power struggles tend to spawn multiple armed groups that seek to influence a peace process or ceasefire negotiations. This can make it hard to distinguish between those groups that should be part of negotiations and thereby providing their consent and those that should be disarmed and reintegrated into civil society, by force or otherwise.

Consent also means that the use of force at the hand of UN peacekeepers must be aligned with the directions of the host nation authorities and the main parties. In practice this often means that UN forces provide various forms of training and logistical support to host nation security forces and support security sector reform efforts that aim to disarm and/ or integrate former armed groups into national security forces. Sometimes, it also includes the conduct of joint or parallel operations against recalcitrant armed groups that are not part of a ceasefire deal or peace agreement. For the UN, this is often done under the pretext of protecting civilians, while for the host nation government it often has more to do with removing potential threats to its post-conflict ruling position, or settling scores from the recent conflict. What makes this extremely challenging for the UN is that many of the national security forces also are responsible for severe human rights violations.

One of the most telling examples is the national army of the DRC, the FARDC (*Forces Armées de la République Démocratique du Congo*), which is regularly reported to be the one of the worst perpetrators of violence against civilians in the DRC (Human Rights Watch 2009). MONUSCO is mandated both to support the FARDC in its operations against armed groups and to protect civilians from physical harm. In March 2013, the Security Council strengthened this dual role for MONUSCO when in order to improve protection of civilians it established a robust FIB 'with the responsibility of neutralizing armed groups' (United Nations 2013b, p. 9). In theory, a mandate to protect civilians will take precedence over any other mandated tasks, such as providing operational support to the FARDC. The 'bedrock' principle of operating under the consent of the host nation, however, makes it untenable for the UN forces to try to stop human rights violations at the hands of the FARDC by force. This would effectively lead to a withdrawal of consent by Congolese authorities. This dual role makes the balancing act of supporting host nation operations while protecting civilians all but impossible for MONUSCO (Holt and Berkman 2006).

This leads us to the principle of *impartiality*, which in theory is supposed to ensure that the UN is seen as a fair and unbiased third actor, and which will help keep the consent of the main parties during UN operations (United Nations 2008c, p. 33). This was a sound principle during the cold war, when UN forces were deployed in the aftermath of inter-state wars, but today is fraught with complications. A main reason for this is that most UN operations are deployed to intra-state conflicts where there is an ongoing competition for power, despite the presence of ceasefire deals or peace agreements. These conflicts seldom see a definitive victory for either side, and, right from their inception, UN operations are therefore either officially or de facto partial in an ongoing armed power struggle.

MONUSCO, for example, is mandated to take sides with the FARDC in its operations against armed groups in the eastern parts of the country (United Nations 2013b). There, the violent power struggle has been on going, on and off, since 1996. The United Nations Mission in the Republic of South Sudan (UNMISS), as another example, is mandated to assist in strengthening the capacity of the government to govern effectively and democratically (United Nations 2011c). This includes improving the security sector, which in practice consists of the armed group which led South Sudan to independence, the Sudan People's Liberation Army (SPLA). As recent developments show, however, the competition for power through armed force is not over in South Sudan, and the SPLA is also responsible for grave violations of human rights (UNMISS 2014).

Adding to this complexity is again the fact that UN operations are set to protect civilians under imminent physical threat. In theory, the UN is supposed to remain impartial with regards to the *wording* of the mandate. From a protection perspective, this would indicate that a UN operation would not remain neutral towards those actors that attack civilians, regardless of who they are. This implies that UN troops may even use military force against the host nation's armed forces to protect civilians. In practice, as the above discussions have shown, direct confrontation will remain untenable, as it would undermine consent and lead to an early exit for UN forces. In practice, the task of protecting civilians is now mostly performed jointly or in parallel with the national armed forces.

Again, returning to our two examples, due to the dismal human rights records of both the FARDC and the SPLA, neither MONUSCO nor UNMISS will ever be perceived as an impartial third actor as long as they are seen as partnering with these forces. In 2010, MONUSCO (then MONUC) attempted to alleviate some of the negative effects of being closely associated with the FARDC by introducing a so-called conditionality policy. Those FARDC battalions that were led by commanders with grave human rights violations on their record would not receive support from MONUC. This policy led the UN to support only about 18 out of a total of 100 Congolese battalions (Kjeksrud and Ravndal 2010). Whether this made the UN operation appear more 'impartial' remains unclear, but in 2013 the concern of being seen as supporting perpetrators of violence against civilians led the UN to develop a 'due-diligence policy for support to host nation security forces' for all UN peacekeeping missions (United Nations 2013a). In 2012, many UNMISS practitioners expressed a need to be more closely involved with SPLA to be able to influence its development.[1] In 2014, however, mandate renewal discussions for UNMISS in the Security Council indicated that most Council members were now 'in favour of the mission suspending state-building activities' (Security Council Report 2014). These examples show that being partial can be both necessary and untenable for UN missions, depending on the prevailing situation on the ground, making the principle of impartiality an unclear concept at best.

Finally, the principle of *minimum use of force* is also withering in current UN operations. In some of the most recent UN mandates, the Security Council has departed significantly from this principle in both wording and intent. The clearest case was the mandate given to the FIB as part of MONUSCO in the Eastern DRC in 2013 (United Nations 2013b). At its inception, the FIB consisted of three infantry battalions (South Africa, Tanzania, and Malawi), one artillery unit and one Special Force and Reconnaissance Company, supported by South African attack helicopters. The Council asked the FIB to 'neutralize armed groups' and thereby reduce the threat to civilians and 'make space for stabilization activities' (United Nations 2013b). Within just a few days during the autumn of 2013, the FIB managed to neutralise the main security threat in the eastern DRC, the armed group M23 (United Nations News Centre 2013). Also in 2013, the UNSC signalled a more aggressive stance for its UN forces in Mali when it mandated them to 'deter threats and take active steps to prevent the return of armed elements' [to key population centres] (United Nations 2013a).

This belief that military force must play a more decisive role to address some of the most challenging threats to peace processes and civilian life may be a welcome development, as nothing has done more damage to the legitimacy of UN peacekeeping operations over the last two decades than their inability to use force when it was needed the most. Still, the UN, alongside most other nations and organisations involved in current conflict resolution and modern warfare, has suffered from 'a deep and abiding confusion about deploying a force versus employing force' (Smith 2006, p. 6). Smith states that there are 'only four things the military could achieve when sent into action in any given political confrontation or conflict: ameliorate, contain, deter or coerce, and destroy' (Smith 2006, p. 323). According to the UN's Capstone Doctrine from 2008, robust UN peacekeeping operations are expected to be able to 'use force at the tactical level ... to defend themselves and their mandate, particularly in situations where the State is unable to provide security and maintain public order', i.e. to protect civilians (United Nations 2008c). Despite the increased importance attached to the task of protecting civilians, UN troops are still left without much guidance as to how that may be done.

A UN military presence based on the principle of minimum use of force will not necessarily influence an armed group's intention to attack civilians, when that indeed is their aim, although some experts claim that there is a correlation between deploying large UN operations and a decrease in civilian casualties (Hultman *et al.* 2013). More convincing is the idea that any military effort to protect must be able to match the will and intent of the perpetrator, based on a thorough understanding of his strategy to attack civilians (Beadle 2014). Historically, UN troops have usually been quite reluctant to confront armed groups directly, and most of their efforts have been to *ameliorate* and *contain* conflict situations at the tactical, or local, level. Somewhat surprisingly, in the Eastern DRC, the FIB managed to *coerce* the M23, by destroying its ability to continue fighting. These functions of force have rarely been linked to the modus operandi of UN peacekeepers, but were indeed able to influence the will and intent of the M23 to give up its campaign. Given the seemingly successful result of the FIB operations, it is quite likely that similar robust approaches will be replicated elsewhere in the future.

The principles of UN peacekeeping are withering because they have remained largely unchanged while the ways wars are fought are changing. Consent, impartiality and the minimum use of force were developed to suit peacekeeping in another era. Current peacekeepers now struggle to make sense of the guidance they (fail to) provide. In future UN operations, Blue Helmets will probably continue to operate with the consent of a host nation but with a clearer understanding that, if the host nation forces target civilians intentionally, the UN forces' responsibility to protect will outweigh concerns over keeping the consent of the host nation. The UN cannot afford to be seen as a close supporter of some of the worst perpetrators of violent human rights violations against civilians. Also, the principle of impartiality may be altered to more clearly reflect that the UN does indeed take sides, with whom it does so and when. Finally, if the Council continues to rely on military force to implement complex mandates, the organisation needs to use force with more utility than is the case today. This must be based on a better understanding of what military force is and what it can and cannot do to move beyond the 'confusion between deploying forces and employing force' (Smith 2006, p. 6).

Witness the return of former troop contributors to UN operations[2]

The chapter moves on to discuss the implications of a return of former troop contributors to UN peacekeeping on the UN's relative force composition and strength. The ISAF exit from Afghanistan may provide an incentive to some of the 'traditional' European UN troop contributors to find their way back to Blue Helmet operations, as the UN will yet again be the dominating mechanism for deploying troops to international conflict resolution efforts. If some of these contributors do indeed return to UN peacekeeping, they will probably aim to provide much needed specialist capabilities. This is based on the observation that major current troop contributors will provide the greatest number of boots on the ground in the future also. First of all, current troop contributors often have a significant number of generalist ground troops available in comparison to the small, highly specialised, and technology-driven European military forces. In addition, for many developing states a large peacekeeping deployment remains a welcome source of income, while for the Europeans larger deployments would involve quite a lot of costs. Finally, this part of the chapter will also ask how a new mix of troops and responsibilities will impact how military force can be utilised in future UN operations.

Earlier, troop contributions to UN operations were a priority for many European states. Ever since the deployment to Afghanistan, however, support for UN peacekeeping from Europeans has mostly come as financial support, not through troop contributions. This could change as ISAF is coming to an end in 2014, freeing European military capabilities. Thus, the lack of any other major NATO operation on the horizon may push some of the ISAF troop contributors to find their way back to Blue Helmet operations, perhaps more due to a sense of obligation than anything else. Troop contributions to UN peacekeeping operations have long been dominated by African and Asian contributors, and more so than ever during the last decade. By October 2011, 37.1 per cent of all military personnel contributing to UN operations were from Africa, 33.9 per cent were from Central and South Asia, while only 9.0 per cent were from Europe (Center on International Cooperation 2012, p. 142). As of April 2014, a total of 97,729 uniformed UN personnel were deployed on four continents to 16 operations, and Table 7.1 below shows the top ten contributors of uniformed personnel.

Table 7.1 Top ten contributors of uniformed personnel (2014). Adapted from UN (2014b)

Troop contributor	Number of uniformed UN personnel
1 India	8,132
2 Bangladesh	8,034
3 Pakistan	8,027
4 Ethiopia	6,628
5 Rwanda	4,709
6 Nigeria	4,614
7 Nepal	4,612
8 Ghana	2,992
9 Senegal	2,967
10 Jordan	2,729

Keeping almost 100,000 peacekeepers deployed currently demands a yearly budget amounting to about $7.83 billion, which is less than half of one per cent of world military expenditures (estimated at $1,747 billion in 2013) (United Nations 2014b). Comparing the list of who contributes peacekeepers with who pays for peacekeeping, however, provides a completely different picture:

There is clearly a massive gap between those that pay for peacekeeping and those who deploy peacekeepers. Some European governments have recognised that a desire to improve UN operations must not only be demonstrated through financial means but must include the will to deploy troops. As one example, Norway's official strategic concept for the Armed Forces from 2009 states that 'the UN's credibility is ... dependent on Western states showing an increased will to take part in UN-led operations than is the case today' (Norwegian Ministry of Defence 2009). As will become evident in what follows below, Norway, alongside all other European states, has so far not been able to live up to that ambition. What follows is a selection of some of the European ISAF-contributors in 2012

Table 7.2 Top ten providers of financial contributions to UN operations. Adapted from UN (2014c)

Financial contributor	*Contribution to UN peacekeeping (percent)*
1 USA	28.38
2 Japan	10.83
3 France	7.22
4 Germany	7.14
5 UK	6.68
6 China	6.64
7 Italy	4.45
8 Russian Federation	3.15
9 Canada	2.98
10 Spain	2.97

Table 7.3 Selected European UN and ISAF contributors (2012). (ISAF 2012; United Nations 2012b)

Troop contributor	*UN*	*ISAF*
1 Italy (20)	1,133	3,816
2 France (24)	1,066	3,308
3 Spain (27)	997	1,481
4 UK (47)	283	9,500
5 Finland (51)	205	176
6 Germany (52)	194	4,900
7 Sweden (71)	63	500
8 Norway (73)	58	525

who also appeared on the UN troop contributor list. Each state's ranking on the UN contributor list is in parentheses.

Judging by these numbers, there is a potential among European ISAF troop contributors post-Afghanistan to fill important functions in UN operations. However, the rudimentary compilation and comparison of these numbers hides a wide variety of political realities and explanations of the motivations of each state concerning when, how and where they wish to send peacekeepers or war-fighters. Instead, what this does indicate is that there is one overarching explanation for the absence of European Blue Helmets in UN operations – the ten-year-long engagement in ISAF, which has put any potential substantial deployment to UN operations on hold. Still, although we are quickly approaching the exit from Afghanistan, it is not possible to discern a broader move among European states which paves the way for future deployments to UN operations. One significant exception, though, is the Netherlands, which has already decided to become a substantial contributor to MINUSMA in Mali (Defence Web 2014). War fatigue and a loss of belief in the chances of success for complex stabilisation/counter-insurgency operations are two possible explanations for the reluctance to engage more closely with UN operations. There are on-going initiatives, though, from influential think tanks and the DPKO, which aim to place this issue higher on the UN's agenda in the near future.[3] Thus, we will probably see more lobbying from the UN Secretariat and others aimed towards convincing potential European troop contributors.

Future European peacekeepers will probably aim to provide much needed specialist capabilities, such as engineers, special forces, strategic airlift, surveillance and intelligence resources. However, integrated UN peace operations have quite a different set-up and approach to operations than the European troops are used to from Afghanistan as well as from former eras of peacekeeping operations. For example, modern integrated missions are led by a civilian Head of Mission, who is also in charge of the military component through the Force Commander. These missions have many integrated civilian components, such as civil affairs, human rights, rule of law, security sector reform and many others. The day-to-day tasks are often guided by comprehensive mission-wide strategic frameworks that involve close coordination with many of the civilian actors. In addition, the most important military task is often to protect civilians from physical harm, within the constraints and principles of the UN's use of force, as discussed earlier in this chapter. Influential troop-contributing countries from the 'south' are now leading the development of military concepts that fit contemporary peacekeeping and urban peace operations. The emergence of peacekeeping training centres in countries like Brazil and India, where broad experience from leading positions in on-going operations inform training and concepts, further amplifies this development. Currently, European countries lack the UN-specific experiences and know-how that significant troop-contributors have gathered over the past decade.

If Europeans return to UN peacekeeping on a broader scale, the future force composition and strengths of UN operations will be very different from what we are seeing today. This may impact the use of military force in future UN operations, as it presents both opportunities and challenges. First of all, many European forces will arrive with recent experience from Afghanistan, where military force has been the most dominant tool in their counter-insurgency campaigns. In a UN setting, they will be met with a civilian-dominated realm of operations, cooperating with troop-contributors that are quite sceptical of a move towards more robust peace-keeping. A pertinent question is whether the Europeans will be prepared to apply more force in defence of the mandate and to protect civilians than seems to be the case today, and what effect this may have on internal UN cooperation. Second, if the Europeans indeed are able to provide much needed specialised capabilities, such as intelligence-gathering and analysis, as will be the main tasks of the Dutch forces in Mali, will this make UN operations a more effective tool, more readily deployed to address complex crises? Answers to these questions will remain elusive for as long as European contributors mostly remain absent from UN peacekeeping, but in order to avoid friction between current and former peacekeepers and to benefit from such a development, it would maybe be wise to think through possible consequences of the latter's potential return to UN operations.

Towards a fragmentation of international conflict resolution

The final section will discuss whether we are moving towards a further fragmentation of responsibilities in international conflict resolution. This could have major impact on the way future UN peacekeepers operate and relate to the use of military force. The operation in Libya reminded us of one important limitation to Blue Helmet operations – the UN itself would never be able to apply force at the strategic level. On that occasion, however, the Council showed that it could respond quickly to a rapidly deteriorating security situation by providing a mandate to protect civilians. Still, the UN will risk becoming marginalised if future conflicts continue to demand military responses that Blue Helmets themselves cannot provide. Currently, in addition to its legal role of providing mandates and supporting regional organisations, the UN keeps deploying a high number of troops to UN-led missions in some of the most difficult operational environments in the world. The UN thus fills an important role by address-ing those conflicts that few other organisations or countries are able or willing to touch. It is not a given that the UN will continue to fill these two functions in the future.

For one, if the Security Council fails to intervene effectively in situations where civilians are threatened with violent human rights violations by their own governments, as has been the case with Syria, it will be a Security

Council that may ultimately outplay its role as guarantor for international peace and security. Earlier cases have indeed shown that when the Council remains deadlocked, other organisations and states have chosen to respond without a mandate from the Council, such as in the case of Kosovo. Future conflicts are likely to produce similar dilemmas. Second, it is not certain that the UN will be able to sustain a similar high number of missions with accompanying personnel in the future. The global financial crisis may influence the organisation's actual ability to deploy and sustain a high number of troops and civilian experts to operations, as nations are cutting back on their defence budgets. It is not a given that deployments to UN operations will be prioritised when nations are met with fiscal constraints. The UN therefore continuously seeks to broaden its cooperation with regional organisations in order to reach functional burden-sharing agreements, such as with the AU, the EU and NATO. The UN is dependent on maintaining good relations with all of these actors to remain relevant, since it obviously cannot address all conflicts on its own, for practical, financial and political reasons. Several factors point towards a future with a more pronounced burden sharing – a fragmentation of responsibilities – between the UN, AU, NATO and EU.

Operations in Somalia, Libya and DRC can provide more substance to this claim. The conflict in Somalia is currently not one that the UN wishes to address with a Blue Helmet operation, although it has tried (and failed) in the past. The African Union Mission in Somalia (AMISOM) is definitely not peacekeeping as envisaged by the UN (Perry 2012). AU troops in Somalia are fighting a large-scale counter-insurgency operation which has led to a high number of casualties on all sides, including civilians. The UN will probably never be able to operate in similar non-permissive operational environments, nor would it be able to apply the necessary amount of military force needed to counter violent opposition similar to al-Shabaab. However, as the MONUSCO FIB has shown, the 'most robust' military approach seen by the UN to date, the UN is sliding towards more forceful responses to threats to the mandate and civilians. The M23, however, was a much less potent enemy than the al-Shabaab. Short of performing peacekeeping operations in Somalia, the UN widely supports AU efforts there, including through a political office and a logistics capacity support package.

Further, the UN has clearly stated that it does not undertake peace enforcement at the strategic level (United Nations 2008c). Thus, when situations like the ones in Libya occur in the future, the UN can at best provide the mandate to protect civilians, and others, such as NATO, must perform the strategic enforcement. The on-going situation in Syria, however, as discussed earlier in the chapter, clearly underlines that it is not necessarily the situation on the ground that decides whether the international community will live up to its responsibility to protect. Although largely deemed a successful effort to protect civilians from physical harm, the operation over Libya underlined that NATO had been generally unprepared for the task of providing direct protection to civilians. Forthcoming revisions of NATO doctrines (non-article V operations) will include guidance on how military force can and cannot be used to protect (Beadle and Kjeksrud 2014). NATO has also been criticised for overstepping its mandate, in that in the end they sought regime change in Libya, not just to protect civilians. In the past, the EU has demonstrated a potential for providing bridging operations to the UN. Operation Artemis in the DRC in 2003 managed to stabilise the areas in and around Bunia town in Ituri, when MONUC was overwhelmed by the communal violence there. Lately, the EU responded to the crisis in the Central African Republic (Premium Times 2014). During the decade in between, however, the organisation repeatedly failed to utilise its Battle Group concept to respond to UN requests for assistance. With both NATO and EU countries suffering from fiscal constraints and deep cuts in defence budgets, the future of strategic peace enforcement operations to protect civilians on behalf of the UNSC remain highly uncertain.

Will the UN manage to maintain its position as an influential actor in peace and stabilisation operations when responsibilities in international conflict resolution fragment? For the time being, the UN will probably continue fill an important niche that other organisations are neither willing nor able to fill. Most UN peacekeepers are deployed to African conflicts that so far have attracted little interest from influential Western states, beyond diplomatic efforts and financial support. With the deployment of French forces to Mali and the recent concern in Western countries for growing Islamic extremism in the Sahel region, this might be about to change. Also, the AU might continue to evolve to take a more assertive role in peacekeeping operations in Africa, challenging the 'monopoly' of the UNSC of deciding when, where and how to intervene. In order to preserve its role as guarantor of international peace and security, the UN will need to clearly define its niche in future peacekeeping so as to not become irrelevant.

Preparing future UN peacekeepers

This chapter has provided a cautious fore*sight* into the near-future development of UN peacekeeping operations. The future presented here suggests a move towards more limited and robust UN peacekeeping mandates, where long term state-building efforts will be less pronounced than shorter term stabilisation and protection efforts through military means. In these future operations, UN troops will actively take sides and sometimes operate without the consent of the main parties of conflict while faced with violent opposition. The operations will be conducted by a mix of Asian and African infantry forces and small numbers of European peacekeepers equipped with new technology and niche capabilities, often alongside regional organisations and influential national states. In order to better manage the uncertainty which comes from these possible developments, future troop contributors could take the following four main points into consideration when preparing future deployments.

First, employing military force will play a more significant role in the implementation of future UN mandates. The emphasis which the UN system puts on protecting civilians from physical harm will demand that future UN peacekeepers are able to use force to protect with more utility than is the case today and not only on the tactical level. This will demand a deeper understanding of what military force can and cannot do to protect civilians from perpetrators that directly target them. This understanding will only emerge through a proper analysis of the strategies and capabilities of different perpetrators of violence (Beadle 2014). Second, the bedrock principles of peacekeeping operations – consent, impartiality and the minimum use of force – must be adapted to the changes in the ways wars are fought. Today, they more often confuse than provide guidance to Blue Helmets on the ground. Future troop contributors should therefore engage the UN in order to clarify what these principles entail in practice. Third, the future force composition will have major impact on the conduct of UN operations. Future troop contributors should make efforts to learn and understand UN peacekeeping through the eyes of the current major troop contributors right now. They have dominated the realm of peacekeeping for the last decade and possess invaluable experience from some of the most challenging conflicts on the African continent. In addition, they will also remain the dominant providers of infantry troops for the foreseeable future. Finally, the on-going fragmentation of responsibilities for international conflict resolution efforts will demand troop contributors that are able to adapt rapidly to different operational frameworks.

How to prepare for the future framework of UN peacekeeping operations, however, is not clear. Today, UN operations are covering a wide spectrum of operations, ranging from traditional monitoring operations, via massive civil–military state-building endeavours, to smaller-scale robust enforcement operations. Somewhat paradoxically, this wide range of operations might in fact be the UN's future niche, as the organisation fills what may be the most important gap in international conflict resolution, by addressing those conflicts that other organisations are neither willing nor able to touch.

Notes

1 Finding based on interviews with UNMISS practitioners during fieldwork in Juba and Bor, South Sudan, November 2012.
2 A big thank you goes to Elise Svarstad at FFI for valuable research support during the production of this section.

3 One example is the 'Providing for Peacekeeping' project, a research project established by the International Peace Institute in collaboration with George Washington and Griffith universities. The project aims to generate and disseminate new knowledge about UN Member States and their approach to UN-led peacekeeping operations in order to: help 'broaden the base' of troop- and police-contributing countries for UN-led operations; improve the quality of troop and police contributions to peacekeeping operations; and fill key capability gaps in UN peacekeeping operations. www.providingforpeacekeeping.org

References

Beadle, Alexander William (2014). *Protection of Civilians – Military Planning Scenarios and Implications.* 2014/00519. Kjeller: Norwegian Defence Research Establishment.

Beadle, Alexander William and Stian Kjeksrud (2014). *Military Planning and Assessment Guide for the Protection of Civilians.* Kjeller: Norwegian Defence Research Establishment.

Carlsson, Ingvar, Sung-Joo Han and Rufus M. Kupolati (1999). *Report of the Independent Inquiry into the Actions of the United Nations during the 1994 Genocide in Rwanda.* New York: United Nations.

Center on International Cooperation (2012). *Annual Review of Global Peace Operations 2012.* New York: Center on International Cooperation.

Defence Web (2014). "First Dutch Troops Head for Mali". April 16. www.defenceweb.co.za/index.php?option=com_content&view=article&id=34411:first-dutch-troops-head-for-mali&catid=56:diplomacy-a-peace&Itemid=111.

Eide, Espen Barth, Anja Therese Kaspersen, Randolph Kent and Karen von Hippel (2005). *Report on Integrated Missions, Practical Perspectives and Recommendations.* Independent Study for the Expanded UN ECHA Core Group. Oslo and London: Norwegian Institute for International Affairs and King's College.

Gowan, Richard (2008). "The Strategic Context: Peacekeeping in Crisis, 2006–08". *International Peacekeeping* 15(4), 453–469.

Gowan, Richard (2009). *The Future of Peacekeeping Operations: Fighting Political Fatigue and Overstretch.* New York: Friedrich Ebert Stiftung.

Gowan, Richard (2012a). "The UN Mission in Syria: Heading for Heroic Failure?" www.worldpoliticsreview.com/articles/11909/the-u-n-mission-in-syria-heading-for-heroic-failure.

Gowan, Richard (2012b). "The Case for Co-Operations in Crisis Management". European Council on Foreign Relations. http://ecfr.eu/page/-/ECFR59_CRISIS_MANAGEMENT_BRIEF_AW.pdf.

Heinecken, Lendy and Rialize Ferreira (2012). "Fighting for Peace: The Experiences of South African Military Personnel in Peace Operations in Africa (Part II)". *African Security Review* 21(2), 42.

Holt, Victoria K. and Tobias C. Berkman (2006). *The Impossible Mandate?: Military Preparedness, the Responsibility to Protect and Modern Peace Operations.* Washington, DC: The Henry L. Stimson Center.

Hultman, Lisa, Jacob Kathman and Megan Shannon (2013). "United Nations Peacekeeping and Civilian Protection in Civil War". *American Journal of Political Science* 54(4), 875–891.

Human Rights Watch (2009). *Soldiers Who Rape, Commanders Who Condone: Sexual Violence and Military Reform in the Democratic Republic of Congo.* New York: Human Rights Watch.

Kjeksrud, Stian (2009). *Matching Robust Ambitions with Robust Action in UN Peace Operations – towards a Conceptual Overstretch?* Kjeller: Norwegian Defence Research Establishment.

Kjeksrud, Stian and Jacob Aasland Ravndal (2010). *Protection of Civilians in Practice: Emerging Lessons from the UN Mission in the DR Congo.* Kjeller: Norwegian Defence Research Establishment.

Norwegian Ministry of Defence (2009). *Capable Force – Strategic Concept for the Norwegian Armed Forces.* Oslo: Norwegian Ministry of Defence.

Perry, Alex (2012). "Somalia's Chance: Can a US-Backed African Force Bring Peace?" *Time.* August 13. http://content.time.com/time/magazine/article/0,9171, 2121096,00.html.

Premium Times (2014). *First EU Forces Deployed in Central African Republic.* www.premiumtimesng.com/news/foreign/158460-first-eu-forces-deployed-central-african-republic.html.

Security Council Report (2014). *South Sudan Briefing : What's In Blue.* www.whatsinblue.org/2014/05/south-sudan-briefing.php.

Smith, Rupert (2006). *The Utility of Force: The Art of War in the Modern World.* London: Penguin Books.

Tardy, Thierry (2007). "The UN and the Use of Force: A Marriage Against Nature". *Security Dialogue* 38(1), 49–70 doi:10.1177/0967010607075972.

Tardy, Thierry (2011). "A Critique of Robust Peacekeeping in Contemporary Peace Operations". *International Peacekeeping* 18(2).

United Nations (1999). *The Fall of Srebrenica: The Report of the Secretary-General.* New York: United Nations.

United Nations (2000). *Comprehensive Review of the Whole Question of Peacekeeping Operations in All Their Aspects (The Brahimi Report).* New York: United Nations.

United Nations (2001). *Implementation of the Recommendations of the Special Committee on Peacekeeping Operations and the Panel on United Nations Peace Operations.* New York: United Nations.

United Nations (2004). *High Level Panel on Threats, Challenges and Change: A More Secure World: Our Shared Responsibility.* New York: United Nations.

United Nations (2005). *World Summit Outcome.* New York: United Nations.

United Nations (2006a). *Peace Operations 2010 Reform Strategy.* New York: United Nations. www.un.org/Depts/dpko/dpko/po2010.pdf.

United Nations (2006b). *Delivering as One: The Report of the Secretary-General's High-Level Panel on System-Wide Coherence.* New York: United Nations.

United Nations (2006c). *Joint Operations Centres and Joint Mission Analysis Centres.* New York: United Nations.

United Nations (2006d). *Note of Guidance on Integrated Missions.* New York: United Nations.

United Nations (2006e). *Integrated Missions Planning Process (IMPP) – Guidelines Endorsed by the Secretary-General on 13 June 2006.* New York: United Nations.

United Nations (2007a). *UN Transitional Strategy Guidance Note.* New York: United Nations.

United Nations (2007b). *Comprehensive Review of the Strategic Military Cell.* New York: United Nations.

United Nations (2008a). *Securing Peace and Development: The Role of the United Nations in Supporting Security Sector Reform.* New York: United Nations.

United Nations (2008b). *Report on the Comprehensive Analysis of the Office of Military Affairs in the Department of Peacekeeping Operations.* New York: United Nations.

United Nations (2008c). *Capstone Doctrine.* New York: United Nations. www.effectivepeacekeeping.org/sites/effectivepeacekeeping.org/files/04/DPKO-DFS_Capstone%20Document.pdf.

United Nations (2009a). *Concept Note on Robust Peacekeeping.* New York: United Nations.

United Nations (2009b). *Protection in Practice – Practical Protection Handbook for Peacekeepers.* New York: MONUC.

United Nations (2009c). *A New Partnership Agenda: Charting a New Horizon for UN Peacekeeping.* UN DPKO and UN DFS. www.un.org/en/peacekeeping/documents/newhorizon.pdf.

United Nations (2010a). *Policy – Civil–Military Coordination in UN Integrated Peacekeeping Missions (UN-CIMIC).* New York: United Nations.

United Nations (2010b). *The New Horizon Initiative: Progress Report No. 1.* New York: United Nations.

United Nations (2010c). *Guidelines – Joint Mission Analysis Centres (JMAC).* New York: United Nations Department of Peacekeeping Operations/Department of Field Support.

United Nations (2010d). *DPKO/DFS Operational Concept on the Protection of Civilians in United Nations Peacekeeping Operations.* New York: United Nations.

United Nations (2010e). *Addressing Conflict-Related Sexual Violence: An Analytical Inventory of Peacekeeping Practice.* New York: United Nations.

United Nations (2011a). *Draft Framework for Drafting Comprehensive Protection of Civilians (POC) Strategies in UN Peacekeeping Operations.* New York: United Nations.

United Nations (2011b). *The New Horizon Initiative: Progress Report No. 2.* New York: United Nations. www.un.org/en/peacekeeping/documents/newhorizon_update02.pdf.

United Nations (2011c). *S/RES/1996.* New York: United Nations.

United Nations (2011d). *S/RES/1990.* New York: United Nations.

United Nations (2012a). *S/RES/2043.* New York: United Nations.

United Nations (2012b). *Ranking of Military and Police Contributions to UN Operations.* New York: United Nations. www.un.org/en/peacekeeping/contributors/2012/july12_2.pdf.

United Nations (2013a). *S/RES/2100.* New York: United Nations.

United Nations (2013b). *S/RES/2098.* New York: United Nations.

United Nations (2014a). *S/RES/2149.* New York: United Nations.

United Nations (2014b). *Ranking of Military and Police Contributions – April 2014.* New York: United Nations.

United Nations (2014c). *Financing Peacekeeping. United Nations Peacekeeping.* www.un.org/en/peacekeeping/operations/financing.shtml.

UN News Centre (2010). "UN Peacekeeping in Consolidation Phase, Says Top Official". www.un.org/apps/news/story.asp?NewsID=35558&Cr=le+roy&Cr1=.

United Nations News Centre (2013). "DR Congo: UN Peacekeeping on Offensive after Defeat of M23, Says Senior UN Official". December 11. www.un.org/apps/news/story.asp?NewsID=46721#.Uz1yMVe5T5M.

UNMISS (2014). *Conflict in South Sudan – A Human Rights Report.* New York: United Nations.

Further Reading

ISAF (2012). "About ISAF: Troop Numbers and Contributions". www.isaf.nato.int/
troop-numbers-and-contributions/index.php.
United Nations (2008). *S/RES/1856.* New York: United Nations.
United Nations (2013). *S/2013/110.* New York: United Nations.
United Nations (2014). *Peacekeeping Fact Sheet April 2014.* www.un.org/en/peace-
keeping/resources/statistics/factsheet.shtml.
United Nations Human Rights Council (2012). *Report of the Independent International
Commission of Inquiry on the Syrian Arab Republic.* New York: United Nations.

8 The new urban operations

Per M. Norheim-Martinsen

As described in Chapter 1 of this volume, the demographic trends are clear. The world is urbanising at an unprecedented rate (Goldstone 2010; UN-HABITAT 2008). More than 50 per cent of the world's population already lives in cities, a share that is expected to grow to 70 per cent by 2050. In sub-Saharan Africa, more than one billion people are expected to live in cities by 2050, while the world's total number of megacities – i.e. cities with more than ten million people – is expected to increase to 25 by 2025 (US Government 2008, p. 23). There is a risk that urbanisation will have a destabilising effect upon many countries. The urbanising states in sub-Saharan Africa in particular are characterised by fast-growing cities, urban sprawl, the growth of slums and massive inequalities between rich and poor. These environments are prone to rising violence and crime. They are also characterised by lack of management and control, which, in turn, may open them up to radical groups, terrorist networks and organised criminals gaining a foothold and being able to operate unchecked. The division between war and crime is getting blurred, as is the division between soldiers and civilians, combatants and criminals. Tomorrow's conflicts are going to move 'out of the mountains' and into the cities, to paraphrase David Kilcullen's latest book (2013). However, as they do so, urban conflicts will, as this chapter argues, take on new traits, challenging what we know about urban warfare from previous campaigns.

The urbanisation trend of the last 20 years has made it increasingly harder to avoid military operations in urban areas, such as in Beirut (1982), Los Angeles (1992), Mogadishu (1993), Gaza (2009), Grozny (1995, 2000), Baghdad (2003) and Fallujah (2004) (Vautravers 2010, p. 439). These developments raise concerns amongst most military officers – and with good reason. The lessons from World War II show that conventional combat operations in urban areas are extremely demanding, given the three-dimensional nature of the urban battleground – the fighting takes place *on, over* and *under*ground. However, the new challenges for military forces are *not* predominantly associated with urban warfare in the traditional understanding of the term. The trend is rather that the speed of urbanisation will have a destabilising effect upon many states. Failed cities may become as central a challenge as failed states are currently. In the future, the danger of vulnerable cities collapsing completely or being subject to massive systemic breakdown as a result of natural disasters, energy shortages, lack of law and order, etc. will have wide security and humanitarian consequences. Tomorrow's cities may contain pockets of heavily armed opposing forces, lawless areas under the control of criminal gangs, but also areas with relative stability, often in the immediate vicinity of each other. As international military operations migrate from predominantly rural to urban conflict environments, military forces will have to carry out tasks such as community policing, disaster management, and more demanding combat operations, simultaneously within the same theatre of operations. This will bring about vast challenges with regard to tactics, doctrine, equipment and training.

To illustrate some of the challenges of the new urban conflicts and ways of dealing with these challenges, this chapter uses a somewhat unfamiliar case. The development of military doctrine and thinking has for a long time predominantly taken place within Western states, which have also naturally dominated the way that, for example, the UN's DPKO thinks about these issues. Although non-Western states have eventually moved up to become the main troop contributors to UN missions, they nonetheless carry out these missions based on Western ideas and concepts. However, in 2004, Brazil took the lead of MINUSTAH, the UN stabilisation mission in Haiti. Although one should be careful not to reduce the 'Haitian problem' to simply one of gang-driven urban conflict, this was nonetheless the overwhelming challenge that Brazil, at the head of the UN peacekeeping force, was faced with in 2004. Hence, rather than comparing the situation in Haiti with the conflicts in the DRC, Lebanon or the Sudans, which have in recent times been home to the more 'typical' UN peacekeeping operations, Haiti bears more resemblance to places like Mexico City, Bogota or Rio de Janeiro (Fishel and Sáenz 2007; Heine and Thompson, 2011). As this chapter argues, the link back to Brazilian security operations in Rio de Janeiro in particular, which has received much attention prior to the 2014 Football World Cup and 2016 Olympic Games, provides a unique source of lessons learned for dealing with challenges of contemporary urban conflict.

For one, MINUSTAH represents one of very few examples of urban peacekeeping. But a case can also be made that in Latin America conflict moved 'out of the mountains' as early as the 1960s, and that contemporary approaches to urban violence in Brazil are in large part influenced by experiences with which Western militaries are unfamiliar. This chapter shows how emerging powers, such as Brazil, are perhaps better equipped to deal with challenges of urban conflict in the future, whereas Western states remain wedded to artificial divisions between war and crime and a built-in fear of urban warfare as something that should be avoided at all cost.

The chapter first aims to distinguish old urban operations from the new. It then moves on to look at Brazilian experiences in dealing with urban security at home and abroad, making a case for the connection between urban security operations in Brazil and Haiti. Recognising that in Latin America urban security has been a major source of concern for decades, the chapter maintains that Western states may have a lot to learn from the Brazilian experience but that the current Brazilian approach to urban insecurity is also much influenced by recent Western experiences with counter-insurgency and peacekeeping.

The new urban operations

Taking the parameters in Chapter 2 as a point of departure, the first part of this chapter seeks to describe some of the traits and challenges that distinguish traditional urban warfare from the new urban operations. In historical terms, World War II marked a turning point for urban warfare. Although siege warfare had been common up until the late 1500s, it gradually gave way to battles between armies in open country. As Alexandre Vautravers describes (2010, pp. 438–439), during World War I, battles were for the most part fought outside the cities from which they took their names, and the front and most of the fighting took place on rough ground in the fields of Western Europe. At the start of World War II, blitz warfare impelled the German Army to move around the cities to avoid getting bogged down. However, the fighting was soon pulled into urban areas as, among other factors, the Allies realised that they were not going to be able to fight the enemy in open ground, and the Soviet scorched-earth strategy inevitably forced the Germans into the cities. Halfway into the war, 'cities in the East and subsequently also in the West gradually became fully fledged targets and the theatres of decisive battles' (Vautravers 2010, p. 439), making the heavy casualties, massive destruction and incredible suffering imposed on the inhabitants of cities such as Stalingrad, Budapest, Arnhem and Berlin earn urban warfare its rightfully bad reputation in the period after.

During the cold war, urban warfare was for the most part avoided, owing in part to the fact that urban centres were considered to be targets of atomic weapons, and troops tended, therefore, to be stationed outside cities. Even if there were clashes in urban areas in this period in places such as Vietnam and Northern Ireland, the need for militaries to fight in urban terrain was largely absent until the global urbanisation trends caught up with events in the 1980s and onward.

Military operations in urban areas, such as in Beirut (1982), Mogadishu (1993), Gaza (2009), Grozny (1995, 2000), Baghdad (2003) and Fallujah (2004), have subsequently been the central source of lessons learned as some states, including the US, Israel, Russia, France and Britain, have developed doctrines, training and equipment for urban warfare. However, a majority of states have not, nor have they had the impetus to do so as, for example, only a handful of states were involved in urban operations in Iraq. Also, much of what is known from past urban operations is, as is argued below, not relevant to the predominantly low-intensity conflict environments we are likely to see in the future. Russian operations in Chechnya in particular (see e.g. Lieven 2001), but also Israeli operations in Gaza and US operations in Iraq, which are often cited as examples of contemporary urban combat, all display strategies and tactics that have immense destructive effects and tend to more or less shut cities down. This will arguably not be possible in tomorrow's megacities.

If we take *mandate* as a point of departure, a first concern is that, although some of the most lethal episodes of armed violence in recent years have taken place in countries such as Mexico and Honduras, none of these countries suffered from conflict or war according to conventional definitions (Briscoe 2013, p. 1). This situation is going to get more blurred as the urbanisation trends continue. As David Kilcullen states, 'Formally declared warfare among nation-states, for example, is likely to keep getting rarer, while violence involving non-state armed groups (whether we call it "war" or "crime") will probably remain the most common and widespread form of conflict' (2013, p. 23). A consequence of this trend will be that the traditional strict division in Western states between internal crime fighting, usually carried out by the police, and external war fighting, usually carried out by the military, will be challenged as situations get out of hand. One pertinent example is the 1992 riots in Los Angeles, which saw the deployment of US National Guard forces alongside state police as local gangs took to the streets following the brutal beating of Rodney King (Delk 2001). But the threshold for employing military force internally in Western states usually remains high. So far, the need to do so has also been absent even if the tendencies are visible in incidents such as the riots in London (2011), Paris (2005) and Malmö (2010). A point to note is that the distinctions between war and crime, military and police tasks are often less clear cut in countries outside the West, where the conflation of military and police tasks tends to be seen, at least from a Western point of view, as a trademark of underdeveloped, less democratic states. Yet, for the challenges posed by extreme cases of urban insecurity, some conflation of tasks may be necessary, although it should be noted that there are obvious dangers in condoning the use of military force, especially in the case of fragile democracies. We shall return to this when we examine the case of Brazil, an emerging power albeit still a young democracy, below.

In an international operations setting, the blurred division between war and crime poses a challenge, as it is hard to define when exactly an urban conflict – whatever we choose to call it – passes the threshold beyond which an international response may come into play. In addition, mandates for intervention are usually given for multinational operations in countries, not cities, even if some operations have taken on more of an urban character than others. UN operation MINUSTAH in Haiti, which we shall return to below, is a notable exception, as operations were mainly concentrated in the capital of Port-au-Prince. Nevertheless, it is feasible to think of future contingencies in which a non-failed state will face the challenge of a failed megacity. The question then is whether it is likely that these states will actually invite a foreign intervention, even if this may be a favourable option as opposed to having the state crack down on its own citizens with military force. Such a situation also raises interesting questions if it should occur in the West, as it very well might. What should authorities do when faced with the choice between breaking the above-mentioned democratic dictum of leaving urban security to the police by deploying military forces and equipping the police so that they can address the problem i.e. between militarising the police and letting things deteriorate until vast urban areas become de facto ungoverned spaces outside state control?

Moving on to *consent*, operations in megacities may come about as the result of negative developments that have taken place over time, but where the causative factor may be, for example, a natural disaster, riots or a massive inflow of refugees, triggering a situation where events spiral out of control. The intervening party may often have to deal with local actors whose willingness to cooperate is limited as there may be divergences between local and central authorities, the latter being ultimately the consenting party to a foreign intervention. However, international forces may only be able to take and keep control of a megacity to a very limited degree over extended periods of time. Nor would they have the local knowledge needed to devise appropriate solutions to immediate security risks while retaining a functioning urban environment for the millions of citizens living there. Extended cooperation and local ownership will be a prerequisite for a successful intervention.

At the same time, the motive for intervening in a failed megacity may be to avoid it becoming a breeding ground for illegal economic activity, terror planning and recruitment etc. Cities are, as Kilcullen (2013, ch. 4) describes them, natural hubs in densely connected global networks, as they are the natural sites for harbours and airports, which connect people with diasporas abroad and allow enormous numbers of international transactions every day. This connectedness is, in turn, exacerbated as information technology becomes available to 'everyone', as described in Chapter 1 of this book. However, in cases where states or groups of states experience lack of government control in a city that leads to unwanted activity spilling over into other countries, national authorities in the states in question may be unwilling to consent to an international presence. The *absence* of a recognisable conflict would then become an obstacle for international intervention.

This suggests that future urban conflicts may be characterised by low *conflict intensity*, although many of the examples of urban conflict seen over recent years – i.e. Gaza 2009, Fallujah 2004, Baghdad 2003 and Aleppo (2012) – seem to suggest otherwise. The latter type of contingencies will continue to pose major challenges for states that have an ambition to retain an urban *warfighting* capability, such as the US, Israel, France, Britain and a few other nations. But, for the majority of states, the challenge will rather be that the more traditional peace and stabilisation operations, which until now have taken place in predominantly rural conflict settings in typically underdeveloped states, will move into the cities. Conflict intensity in these contingencies will, as opposed to in the examples above, be generally low.

In terms of *relative force composition and strength*, the opponents will consist of relatively small forces with limited military resources. But the possibility of mobilising large crowds of people in a short space of time will make sure that all tactical operations will be subject to limited oversight and the risk of quick escalation. The battle of Mogadishu (1993) between US forces and Somali militias, which featured in the box office movie *Black Hawk Down*, is a timely example of how quickly – and disastrously – situations may escalate in urban environments. Yet again the main challenge may not necessarily be armed militias under the leadership of war lords, fighting for control of cities, as briefly discussed by Tore Nyhamar in the next chapter. Rather, the challenge may be more or less organised criminal networks under the leadership of crime lords who profit from the lack of government control. In these situations, the divisions between militias, terrorist organisations, criminals who may or may not act as community leaders and even corrupt representatives of authority, will be blurred. The aim of an operation will often be to re-establish government control, but there will be many potential spoilers and motives for preventing it.

Finally, even if these types of operation may offer military challenges at the lower end of the scale, the city's *operational environment* will always grant the defender, who will have intimate knowledge of his surroundings, the advantage over the intervening force. Military officers' well-known fear of urban combat stems from their knowledge of exactly how demanding an operational environment the city is. Battle takes place in three dimensions: *on* the ground; *under*ground in basements, subways, tunnels and sewage systems; and *over* ground in upper floors of buildings and on rooftops. In addition, there is a need to be aware of structures, such as electricity grids and networks of communication, knowledge that is of less importance in rural environments and which is not necessarily to be found within the military. There are usually many civilians present, and battle ranges are short. Streets, corners and other physical barriers tend to constrain and channel advancing troop movements, while defenders will always have ample opportunities for concealment and cover. That said, there are ways of negating the disadvantages that the cityscape poses to an advancing force.

US experiences in Baghdad and Fallujah in particular have fed in to a renewed interest in and debate about urban operations in online forums, such as *Small Wars Journal*. Israel too has been at the forefront of thinking on urban operations, fuelled by experiences in Beirut (1982) and later Gaza, and has been known for innovative approaches to urban warfare. In Nablus in 2002, for example, Israel employed a strategy known as 'walking through walls' (alternatively 'infestation' or 'inverse geometry') in which:

> the Israeli Defence Force (IDF) soldiers used none of the streets, roads, alleys or courtyards that constitute the city, and none of the external doors, internal stairwells and windows that constitute the order of buildings, but rather moved horizontally through walls, and vertically through holes blasted in ceilings and floors.
>
> (Shunk 2014)

The use of helicopters and bulldozers to clear barriers, avoid being channelled into areas where the opponent is strong and create new and unexpected directions of attack has been a trademark IDF strategy since operations in Beirut (1982). However, the 'walking through walls' strategy, as has been pointed out, signals a fundamental shift from thinking about the city as merely a particularly challenging operational theatre of war to a conception of the city as the very medium of war (Shunk 2014; Weizman 2006). This shift of thinking is also evident in ideas about urban metabolism and conceptions of the city as a system or an organism that needs to be understood and managed to devise appropriate interventions (see Kilcullen 2013, 41–45). Kilcullen is, for example, critical of the US urban operations in Iraq, as their success involved basically 'putting the city on life support'. He is also critical of the IDF strategy just described, because it involves massive damage to the fabric of the city – and, indeed, to innocent civilians as they experience whole squads of soldiers literally blasting their way through their living rooms (Kilcullen 2013, 109–111).

These examples go to illustrate the immense destructive effect military operations tend to have on cities as they are turned into theatres of war. But as we have seen so far, future urban conflicts are likely to take place in cities that are not, and should not be turned into, warzones (see Graham, 2010). The question is whether this implies that there is no 'utility of force' in dealing with the problems of future urban conflicts. There are simply very limited experiences on which to build to answer this question, at least if we limit our search to the 'usual suspects'. When broadening the search, however, it is tempting to point out that, in a place such as Latin America, the fighting moved 'out of the mountains' and into the cities years ago. We shall, therefore, shift the focus to Brazil, an emerging power in several respects, including an emerging lead nation in UN peacekeeping operations (Kenkel 2010).

Urban peacekeeping – the case of Brazil

In Brazil, as in many Latin American countries, urban violence has been a source of concern for decades (Briscoe 2013). Violence in Brazil reflects structural problems of inequality of citizenship and opportunity. But it has been exacerbated by the emergence of organised violent groups, which have become powerful enough to wrest control of significant parts of the urban landscape – *favelas* – from the state. The popular conceptions of Brazil's favelas tend to depict these areas as places of extreme violence, as seen in movies like *City of God* and *Tropa de Elite*. While these informal urban communities have for much of their history been poor yet peaceful places, the lack of state presence has seen an expansion of violence as organised criminal groups emerged following Brazil's return to democracy. The limits of formal recognition and basic services and the near-absence of rule of law in many places enabled the expansion of violent 'parallel states' and other informal forms of governance by criminal groups. In light of the renewed attention brought by the World Cup in 2014 and Olympics in 2016 and recognising that the authority of the state has been slowly ceded to illegal actors, space has emerged for discussions about new approaches to public security in Brazil. It may be argued that the idea of the 'barbarians one day coming down the hills', which permeates the collective psyche in Rio, has prompted support for these more forceful tactics to improve public security ahead of events in 2014 and 2016. But these developments are also linked to the shift in Brazilian policy towards participating in international peacekeeping operations since around 2004.

As Jorge Zaverucha notes, the armed forces have always had a role in the internal security of the state in Brazil (Zaverucha 2000). But the external utility of military force has been negligible, since Brazil is surrounded by friendly states and has not fought a major war since 1870. However, Brazil has always been a strong supporter of the UN and has contributed military forces to various missions. Its participation has been relatively small and focused on Chapter VI missions until 2004, when Brazil agreed to take the lead of MINUSTAH, the UN stabilisation mission in Haiti. In a Chapter VII enforcement operation, Brazil assumed the command of a robust military force consisting of roughly 10,000 personnel, including 7,200 military troops, of which 1,900 are Brazilian, and 2,800 Formed Police Units. This shift in policy was largely due to President Luiz Inacio Lula da Silva's drive to translate economic clout into a global political role for Brazil (Hirst 2007; Mendelson and Forman 2011). Brazil did not assume a link to domestic security, at least not as a principal cause for taking part in the operation, even if some commentators have suggested that a reason for participating in MINUSTAH was to use the operation as a training ground to prepare the Brazilian Armed Forces for use in public security emergencies back home (see e.g. Diniz 2007). However, one cannot escape the fact that Haiti represented, at least at the outset, a somewhat familiar challenge for Brazil, a point that is often raised by Western commentators who tend to assume that a Latin American lead nation would offer a Latin American solution to a typically Latin American problem of urban violence.

Variously characterised as a nightmare, predator, collapsed, failed, failing, kleptocratic, phantom, virtual or pariah state, one should be careful not to reduce the 'Haitian problem' simply to one of 'gangs with guns' (Buss 2008; Pace and Luzincourt 2009). Yet this was the overwhelming challenge that Brazil at the head of the UN peacekeeping force was faced with as it intervened in 2004. Brazilians will, for obvious reasons, object to any comparisons being made between Haitian and Brazilian society. Nonetheless, there are both obvious and more tacit connections between urban violence – and ways of dealing with it – in Brazil and in Haiti that need to be fleshed out to afford a better understanding of the way that urban security strategies have evolved in Brazilian cities, and Rio de Janeiro in particular, and eventually inside the UN.

First, there are connections between how to conceptualise the situation on the ground in parts of Rio de Janeiro and Port-au-Prince respectively. Both have been marked by the emergence of so-called ungoverned spaces, where people live their lives with almost no state interference or engagement. Although it might be possible for many of the individuals living in areas lacking state presence to lead reasonably normal lives, they tend to lack opportunities available to those outside the favelas or shantytowns, such as proper healthcare and education. The limited opportunities available to marginalised urban residents often pushes many, particularly young men, into drug-related economic activities, with insecurity, imprisonment and early and violent deaths frequently a result. These marginal urban communities are places 'where the murder rate is seen as symptomatic of societies where pervasive political tensions are played out at the individual and small group level rather than being mobilized as insurgencies' (Pace and Luzincourt 2009, 8). In other words, violence does not take the form of politicised national movements but usually occurs or is confined to limited geographical spaces, despite the fact that levels of violence may surpass those in many insurgencies or wars. Authors such as David Kilkullen (2013) describe these situations as post-counterinsurgency, where a fusion of criminal and political forms of violence occurs in increasingly ungoverned, crowded and connected urban spaces.

Second, there is the actual chronology of events. In Haiti, violence escalated throughout 2004 and 2005, and in 2006 the Brazilian battalion was given the green light to intervene militarily and forcefully in gang strongholds. The operations were designed to regain territorial control and insert a permanent presence within the areas controlled by the gangs, thus relieving them of their power base (Chagas 2010; Dorn 2009; Dziedzic and Perito 2008). By July 2007, MINUSTAH and the Haitian authorities had regained control of all sections of the capital, and rates of violent crime dropped. Several hundred gang members were apprehended, and MINSTAH established a permanent presence in notorious neighbourhoods like Cité Soleil and Bel Air, which had previously been inaccessible to local and international authorities. From a military point of view, the operations were a resounding success for the Brazilians, who made use of tactics such as the establishment of strong points inside gang controlled areas; use of night-time operations to reduce civilian casualties; extensive use of intelligence in all phases of the operations; and implementation of Quick Impact Projects to repair damage caused by MINUSTAH and win the confidence of the population. This is not to say that the heavy-handed way in which the operations were carried out did not raise controversy. One contingent reportedly fired some 60,000 rounds in what might only be described as outright street battles, sometimes using snipers to 'take out' gang members. However, using the strong points as a base from which to carry out hearts and minds projects and gather information, Brazilian forces were able to calm the situation and gain the trust of the local population, such that in the end, as one Brazilian officer put it, 'they were handing the gangsters over to us' (interviews with Brazilian Armed Forces personnel, Brasilia, September 2012).

The operations carried out by Brazilian MINUSTAH troops between 2006 and 2007 subsequently became a source of inspiration for the so-called 'pacification strategy' which was introduced to take back state control in the favelas of Rio de Janeiro in 2008. The strategy has involved three phases: (i) a *tactical intervention* by the military police – the Special Police Operations Battalion, or BOPE (*Batalhão de Operacões Policiais Especiais*) – with support of the Army or the Marine Corps, to recoup areas controlled by armed groups; (ii) a *stabilisation* phase designed to secure and calm these areas; and (iii) a *consolidation* phase involving permanent deployment of specially trained Pacification Police Units (*Unidade de Policia Pacificadora*) or UPPs (Muggah and Mulli 2012). There are clear connections between MINUSTAH and urban engagement in Rio de Janeiro, as troops from the Brazilian Marine Corps who participated in the first two 'pacifications' of the favelas of *Vila Cruzeiro* and *Complexo do Alemão* in November 2008 were part of the MINUSTAH offensive against the gangs in Port-au-Prince in 2006–2007. The MINUSTAH strong points were also the model for the UPP concept, and UPP units even resemble UN peacekeepers with similar army fatigues and blue caps. Also of particular interest is that the term 'peacekeeping', even if it does sit somewhat uneasily with Brazilian police and army personnel, is reportedly used inside the Rio police to describe their current operations in the city's favelas (interviews with police officers, Rio de Janeiro, June 2012).

Again, the close relationship between the Brazilian Armed Forces and the police, both in terms of thinking and organisation, must be emphasised (Zaverucha 2000). Inside this special relationship, lessons travel both ways. For example, the Brazilian Armed Forces regularly train with the military police, i.e. BOPE, as part of preparations for deployment to Haiti. BOPE, in turn, traces its origins back to the army's Special Operations Forces. BOPE's early interventions in the Rio favelas were also typically quick in-and-out raids in the shape of 'aggressive incursions complete with armoured vehicles known colloquially as *caveirãos* ("big sculls")' (Muggah and Mulli 2012, p. 63).[1] However, as 'pacification' has required BOPE to retain a more visible presence for a longer period of time while awaiting relief from the UPPs, they have also had to revise their traditional aggressive style of engagement and move towards an approach more in line with the hearts and minds thinking in contemporary counter-insurgency operations (interviews with police officers, Rio de Janeiro, June 2012). This need to adapt military approaches to differing operational circumstances, rather than focusing strictly on established counter-insurgency lessons and principles, is discussed in depth by Robert Egnell and David Ucko in Chapter 4 of this volume.

In terms of improving Rio's public image in the run-up to the 2014 World Cup and the 2016 Olympics, the 'pacification' strategy could be seen as a success. More than 30 favelas have been pacified. However, the 'pacification' strategy has been criticised from a number of quarters. The participatory nature of community policing approaches was not translated over to the UPPs and was instead executed based on top-down government approaches. It is held that the 'pacification' strategy has been more concerned with re-establishing the sovereignty of the state in contested territories than with rebuilding trust between police and communities. Even if the role of the military has been in principle limited to supporting the police in the first phase of the 'pacification', stipulated to last from 100 to 150 days, the Army retained a sizeable force in *Alemão* four years after they entered it. The approach has also been criticised for being a developmental project designed to bring desirable urban areas under the control of the state for speculation and investment purposes rather than being concerned with the welfare of favela dwellers (see e.g. Braathen *et al.* 2012). There are signs that 'pacification' is contributing towards rocketing real estate prices, forcing residents to migrate to other favelas (Muggah and Mulli 2012, p. 66). Some critics have also raised concerns about the altering of social relations in the favelas, as the UPPs move in to fill the power vacuum left behind by the gangs, and the danger of creating 'quasi police states' (Mulli 2011).

Similar concerns have been raised with regard to the heavy international military involvement in Haiti and the ability to re-establish a sense of normalcy in a society as long as the military presence remains. Although the focus of MINUSTAH is shifting towards enabling the Haiti National Police and strengthening the UN Police Mission, Brazilian authorities have so far insisted on keeping its 1,900 soldiers in place to contain any re-escalation of violence (Norheim-Martinsen 2012). As such, the militarised approach to urban violence may seem to reflect a particular mind-set or emerging set of norms for the management of violence in Brazil. An interesting question is whether this represents a new turn or merely the continuation of previous militarised forms of policing inside Brazil, which are now also being exported to places like Haiti.

Urban conflict and utility of force

The Rio 'pacification' strategy has been framed both in terms of *counterinsurgency* and *community policing*. However, rather than putting the right label on it, it is important to understand what 'pacification' actually entails, since it has already been suggested as a possible 'model to the region and the world' (Kolbe and Muggah 2012). Similar recommendations have followed in the wake of MINUSTAH, including a report by the United States Institute of Peace which advises that 'the lessons from the resounding success achieved in Haiti should be captured and put into practice wherever missions are challenged by illegal armed groups' (Dziedzic and Perito 2008).

First of all, it is necessary to recognise that both approaches have a clear military component, when it comes to the Brazilian Armed Forces' actual participation in the operations, the tactics employed, and the overall way of thinking. In the words of one commentator: 'Through their occupation of spaces once governed by armed criminals, the pacification police aspire to bring peace through the metaphors of war' (Tierney 2012, p. 7). In fact, the situations in Port-au-Prince and many Brazilian cities all fall rather neatly into what Rupert Smith dubs 'wars amongst the people'—wars where 'civilians are the targets, objectives to be won, as much as an opposing force' (Smith 2005, p. 4). These new modes of urban conflict might be argued by some to be the quintessential 'war amongst the people' with all the characteristics and challenges that they pose, such as difficulties separating combatants from civilians, shifting levels of conflict intensity, uncertainties about jurisdiction, difficult topographical and human terrain and a constant struggle to win hearts and minds. In this sense, while the situations in places like Port-au-Prince and Rio can rightly be regarded as demonstrating chronic urban insecurity, they might also be similar to what others consider as 'new urban wars' and represent a new paradigm to which armies all over the world are being forced to adapt (see Kilcullen 2013).

Against this backdrop, it would be wrong to view the 'pacification' strategy as a *re*militarisation of favela policing. Rather, it represents a new form of security operations, influenced by Brazil's experiences in Haiti but also by general lessons learned from military operations in places such as the Balkans, Iraq and Afghanistan. It represents a militarised approach to urban security but a new and 'softer' kind of militarised approach. On the one hand, UPPs carry out operations that resemble military peacekeeping or counter-insurgency, the latter a concept or approach discussed in depth by Egnell and Ucko in Chapter 4. On the other, the UPPs incorporate ideas of community policing and lessons learned from previous (often failed) attempts to reform the police (Tierney 2012, pp. 30–40). The current approach is undoubtedly a step forward from the era of extrajudicial killings and brutal incursions by the army and the military police in Brazil, but the military way of thinking is still visible – if not dominant – in the way that urban security is dealt with in Rio.

At the same time, there is a need to appreciate the fact that, in the case of Rio de Janeiro, as in other cities in Brazil and beyond, there is in fact, to paraphrase Rupert Smith again, 'utility of force' insofar as regaining territorial control is a precondition for engaging with the local community at all. Also, by involving the military to demonstrate overwhelming shows of force in the first phase of 'pacification', the authorities may in fact have prevented serious clashes between law enforcers and favela dwellers in subsequent 'pacifications'. After *Complexo do Alemão*, the largest favela in Rio and known for being the centre of the drug trade, was conquered in less than two hours, the message became clear that any future resistance would be futile. Subsequent 'pacifications' have gone by largely without serious incidents.

Yet that there may be utility of force in tackling urban security is an argument that sits uneasily with Western states, for whom crime and violence is an issue that should be tackled by the police and not the armed forces. This line of reasoning is also present in the calls for reducing the military presence in Haiti in favour of a heavier role for the police. Yet the levels of violence in Rio, Port-au-Prince and several other Latin American and Caribbean cities tend to surpass those of many conventional conflicts. There is a danger, therefore, that Western models or ideas of policing and security are not suited to meet the challenges raised by urban violence on the scale seen here. Long-term strategies clearly need to include a gradual transition to the police and long-term community projects designed to offer social security, employment, political participation and other opportunities, rights and services, as discussed by Guro Lien in Chapter 5. However, it might also be that Western states need to accept that there may be utility of force in situations where urban violence exceeds certain levels. This union of military operations and community policing approaches is already being acknowledged in post-counterinsurgency discussions (Kilcullen 2013), and is interestingly being incorporated into policing operations when dealing with gang-related violence in large US cities (Bertetto 2013).

Conclusion

This chapter has argued that many of the lessons of traditional urban warfare will not be relevant for future urban operations. That is, the urbanisation trends of the last 20 years suggest that urban warfare in the more traditional understanding of the term will continue to pose challenges for states that have an ambition to retain an urban *warfighting* capability, such as the US, Israel, France, Britain and a few other nations. Yet for the majority of states the challenge will rather be that the more traditional peace and stabilisation operations will move from rural to urban conflict settings. As they do, we need to revisit other examples of urban conflict and ways of dealing with them than those that have been the source of attention in the literature on urban military operations so far.

In the future, it is unlikely that we will avoid military operations in urban environments, which has been the default 'strategy' for most Western militaries for years. This is where the fighting will inevitably take place – or where the need for foreign intervention to alleviate human suffering or prevent new threats from materialising will be. But, in adapting military doctrines and capabilities to this new reality, it is absolutely vital to avoid the destructive effect military operations tend to have on cities as they are turned into theatres of war. Future urban conflicts are likely to take place in cities that are not, and should not be turned into, warzones (see Graham 2010). The question is whether this implies that there is no 'utility of force' in dealing with the problems of future urban conflict. The Brazilian experiences from Port-au-Prince and Rio de Janeiro offer little clarity but some indications as to the shape of things to come.

The Brazilian case study connects long-established issues of dealing with urban violence in Latin America with ongoing debates in the US and beyond about post-counterinsurgency approaches in increasingly urban conflict settings. The lessons from the Brazilian experience clearly have implications in broader contexts. Post-counterinsurgency approaches emphasise the city and urban environments as increasingly important arenas for security. Whether or not military engagement will look like that of Brazilian forces in the favelas of Rio de Janeiro or in Port-au-Prince under MINUSTAH is unclear. However, in these contexts there are signs that the army as an institution has found utility of force in an organised and accountable manner in situations of extreme urban insecurity. While this is of clear interest to military actors, modified approaches applied to civilian police forces may yield benefits. Indeed, there is also an increasing recognition that law enforcement in the US and Europe may benefit from taking on 'adapted counter-insurgency' approaches that resemble the UPP and community policing approaches in Rio de Janeiro when dealing with criminal gangs in large cities. However, these developments raise concerns among some who see the proliferation of SWAT teams and the use of 'paramilitary tactics' by the police as a worrying trend (Balko 2013; Graham 2010; *The Economist* 2014). But principled resistance to using the military or militarised approaches may also lead to unnecessary human suffering, crime and risk of conflict escalation in cases where there is utility of force.

Again we see the tension between military and civilian solutions to situations that are hard to place along a traditional crime–conflict continuum. In Brazil, it is in the transition from military occupation to community policing that something is currently lacking and where the lines of responsibility between the military and the police start to blur. The division of labour between the military and civilian instruments of the state is sharp in most Western states. This is reflected in discussions about conflating these instruments in contemporary international operations like Afghanistan, where concepts such as *Comprehensive Approach* and *Whole of Government* have evolved from a focus on integration to a tacit agreement that the military should stick to providing security and leave the state building to others (Norheim-Martinsen 2013).

In Brazil, on the other hand, the conflation of issues like policing, peacekeeping, humanitarian aid and social development is seen as unproblematic, or even as an asset. Indeed the motto of the Brazilian Army is '*Braço Forte, Mão Amiga*' ('the strong arm and the giving hand'), signalling the multiple roles that the Army does and should have once installed in a community. The Army is mostly seen as managing these multiple roles well, generally being regarded as one of the institutions in society that the Brazilians trust the most. The police, on the other hand, are generally seen as brutal and corrupt. What can be seen as the remilitarisation of ways of dealing with urban security in Brazil may therefore be understandable and even welcome from a pragmatic point of view. The change in approach was motivated by a confluence of several factors, which created the capacity and need to address social problems in Brazil, and a group of actors which had the will and trust of others actors to enable this transition. Nevertheless, from a lessons learned perspective it needs to be mentioned that there are inherent dangers in romanticising the use of force in matters of internal security, especially in a young democracy such as Brazil, since there may be temptations on the part of federal and state authorities and the Armed Forces themselves to have the Army take on ever more tasks. Then again, the hybrid nature of future urban conflict seems to suggest that solutions are to be found *in between* civilian law enforcement and military operations. This suggests that civil–military cooperation needs to improve and that the clear division of tasks in Western states – which is there for perfectly good reasons – may have to be rethought when preparing for future urban operations.

Finally, as a new lead nation in the UN, Brazil has been able to push along much needed changes in the way that, for example, the UN handles intelligence (Dorn 2009; Norheim-Martinsen and Ravndal 2011). Good intelligence was undoubtedly a key factor behind the success of the Brazilian operations against the gangs in Haiti in 2006–2007, and the Joint Mission Analysis Centre, one of the first to be set up in MINUSTAH, is now a mandatory component in all UN missions. Another key factor was the people-centric approach, which involved the use of nighttime operations to reduce civilian casualties and the establishment of strong points inside gang-controlled areas from which to launch Quick Impact Projects to repair damage and reach out to the population. Notably, the establishment of the strong points necessitated robust use of force on a tactical level, as it engaged the gangs head-on in areas under their control, but the underlying rationale was always to reach out to improve conditions for the normal citizens of Port-au-Prince. As remarked by Alexandre Vautravers (2010, p. 442): 'The most salient feature of a city is its population.... Armed forces must take into account the vulnerability and expectations of the civilians at all stages of their planning, communication, intelligence and logistics operations'. In this respect, cities are no different from other conflict environments but are essentially, again in the words of Rupert Smith, 'wars amongst the people'.

Note

1 The popular movies *Tropa de Elite I and II*, which feature BOPE, are also regarded as giving a fairly truthful picture of this early phase.

References

Balko, R. (2013). *Rise of the Warrior Cop: The Militarization of America's Police Forces*. New York: PublicAffairs.

Bertetto, J. A. (2013). "Counter-Gang Strategy: Adapted COIN in Policing Criminal Street Gangs". *Small Wars Journal*, 9(11).

Braathen, E., Bartholl, T., Christovão, A. C. and Pinheiro, V. (2012). *WP3 Settlement Field Report: Rio de Janeiro*. Oslo: Norwegian Institute for Urban and Regional Research, NIBR.

Briscoe, I. (2013). *Non-conventional Armed Violence and Non-state Actors: Challenges for Mediation and Humanitarian Action*. NOREF Report, May. Oslo: Norwegian Peacebuilding Resource Centre.

Buss, T. F. (2008). *Haiti in the Balance: Why Foreign Aid Has Failed and What We Can Do About It*. Washington DC: Brookings Institution Press.

Chagas, C. V. B. (2010). "MINUSTAH and the Security Environment in Haiti: Brazil and South American Cooperation in the Field". *International Peacekeeping* 17(5), 711–722.

Delk, J. D. (2001). "The Los Angeles Riots of 1992". In M. C. Desch, Ed. *Soldiers in Cities: Military Operations in Urban Terrain*. Carlisle, Pa: The Strategic Studies Institute, US Army War College.

Diniz, E. (2007). "Brazil: Peacekeeping and the Evolution of Foreign Policy". In J. T. Fishel and A. Sáenz, Eds. *Capacity Building for Peacekeeping: The Case of Haiti* Washington, DC: National Defense University Press, 91–111.

Dorn, A. W. (2009). "Intelligence-led Peacekeeping: The United Nations Stabilization Mission in Haiti (MINUSTAH, 2006–2007)". *International Peacekeeping* 24(6), 805–835.

Dziedzic, M. and Perito, R. (2008). *Haiti: Confronting the Gangs of Port-au-Prince.* Special Report 208, September. Washington, DC: United States Institute of Peace.

The *Economist* (2014). "Paramilitary Police: Cops or Soldiers?" 22 March.

Fishel, J. T. and Sáenz, A., Eds. (2007). *Capacity Building for Peacekeeping: The Case of Haiti.* Washington, DC: National Defense University Press.

Goldstone, J. A. (2010). "The New Population Bomb". *Foreign Affairs* 89(1), 31–43.

Graham, S. (2010). *Cities under Siege: The New Military Urbanism.* London: Verso.

Heine, J. and Thompson, A. S., Eds. (2011). *Fixing Haiti: MINUSTAH and beyond.* New York: United Nations University Press.

Hirst, M. (2007). *South American Intervention in Haiti.* Comment, April. Madrid: FRIDE.

Kenkel, K. M. (2010). "South America's Emerging Power: Brazil as Peacekeeper". *International Peacekeeping* 17(5), 644–661.

Kilcullen, D. (2013). *Out of the Mountains: The Coming Age of the Urban Guerrilla.* New York: Oxford University Press.

Kolbe, A. R. and Muggah, R. (2012). *Haiti's Urban Crime Wave? Results from Monthly Household Surveys.* Strategic Brief, March. Rio de Janeiro: Instituto Igarape.

Lieven, A. (2001). "Lessons of the War in Chchnya, 1994–1996". In M. C. Desch, Ed. *Soldiers in Cities: Military Operations in Urban Terrain.* Carlisle, Pa: The Strategic Studies Institute, US Army War College, 57–74.

Mendelson Forman, J. (2011). "Latin American Peacekeeping: A new era of regional cooperation". In J. Heine and A. S. Thompson, Eds. *Fixing Haiti: MINUSTAH and beyond.* New York: United Nations University Press.

Muggah, R. and Mulli, A. S. (2012). "Rio Tries Counterinsurgency". *Current History* 111(742), 62–66.

Mulli, A. S. (2011). *Patrolling Rio's Favelas – From Pacification to Police State?* Zurich: ISN, Center for Security Studies.

Norheim-Martinsen, P. M. (2012). *Brazil: An Emerging Peacekeeping Actor.* NOREF Report, November. Oslo: Norwegian Peacebuilding Resource Centre.

Norheim-Martinsen, P. M. (2013). *The European Union and Military Force: Governance and Strategy.* Cambridge: Cambridge University Press.

Norheim-Martinsen, P. M. and Ravndal, J. A. (2011). "Towards Intelligence-Driven Peace Operations? The Evolution of UN and EU Intelligence Structures". *International Peacekeeping* 18(4), 454–467.

Pace, M. and Luzincourt, K. (2009). "Haiti's Fragile Peace: A Case Study of the Cumulative Impacts of Peace Practice". *CDA Collaborative Learning Projects:* Reflecting on Peace Practice Project, Case Study Report, November 2009.

Shunk, D. (2014). "Mega Cities, Ungoverned Areas, and the Challenge of Urban Combat Operations". In *2030–2040 Small Wars Journal,* 23 January (http://smallwarsjournal.com/jrnl/art/mega-cities-ungoverned-areas-and-the-challenge-of-army-urban-combat-operations-in-2030–2040).

Smith, R. (2005). *The Utility of Force: The Art of War in the Modern World.* London: Penguin Books.

Tierney, J. (2012). "Peace Through the Metaphor of War: From Police Pacification to Governance Transformation in Rio de Janeiro". MA Master thesis, Massachusetts Institute of Technology, Boston.

UN-HABITAT. (2008). *State of the World's Cities 2008–2009: Harmonious Cities.* New York: United Nations Human Settlements Programme.

US Government. (2008). *Global Trends 2025: A Transformed World.* Washington DC: National Intelligence Council.

Vautravers, A. (2010). "Military Operations in Urban Areas". *International Review of the Red Cross* 92(878), 437–452.

Weizman, E. (2006). "Lethal Theory". *LOG Magazine,* April.

Zaverucha, J. (2000). "Fragile Democracy and the Militarization of Public Safety in Brazil". *Latin American Perspectives,* 27(3), 8–31.

9 Transnational operations

Tore Nyhamar

Today, conflict does not know borders. The transnational dimension is indeed characteristic in the two largest international operations during the last decade, in DRC and Afghanistan. Richard Holbrook, who made known the concept of AfPak (Afghanistan and Pakistan as a single theatre of operations), explained the need for it in this way:

> It is an attempt to indicate and imprint in our DNA (sic) the fact that there is one theatre of war, straddling an ill-defined border, the Durand Line, and that on the western side of that border, NATO and other forces are able to operate.
>
> (William Saffire, "World Wide Web of Words".
> *New York Times* (2009), p. 10)

This chapter will explore how the fact that conflict is becoming increasingly transnational affects an international force. To find utility of force, it will be necessary to find new institutional and political mechanisms in addition to or as a substitute for increasing the area of operations. Ideally, in a truly transnational *operation*, there is not only one theatre of war, but also one theatre of operations. However, for political and practical reasons this is often not feasible. Therefore, the chapter will explore new institutional and political mechanisms that can substitute for or supplement increasing the area of operations.

Handling transnational conflict matters for two reasons. First, conflict taking place in the territory of more than one state is the rule rather than the exception. Fifty-five per cent of all rebel groups active since 1945 have used territory outside their main or target state (Salehyan *et al.* 2011). Conflicts are increasingly affected by transnational factors (Checkel 2013a). Second, it matters for the outcome of conflict. ISAF and MONUSCO are both hampered in their efforts by groups operating on both sides of the border. 'Border control' was achieved in all eight counter-insurgency wins on record after 1945 (Paul *et al.* 2011, Appendix p. xx). More conflicts than previously are transnational and it is a strong influence on the outcome of the conflict. The trend towards transnational operations is thus problem-driven.

The increase in transnational conflict is explained by the economic trends described in Chapter 1. Most of the future world economic growth will take place outside the West. The world is becoming one economic system, with even the most remote areas connected to world markets, be it an Afghan peasant growing poppy or a labourer digging for diamonds in DRC. Technology connects people in other ways, too. Africa is increasingly connected by mobile phones. Ideas and information are spread, enabling people to mobilise, increasingly across boundaries, in support of their ideas. The trend towards increased interconnectedness will continue, fuelling more transnational conflict in the future.

The remainder of this chapter discusses how transnational operations can be understood in terms of the conflict parameters: mandate, conflict intensity, operational environment, consent, and relative force composition and strength. The chapter is divided into three main parts. The first part examines what the transnational dynamics of conflict are. Taking the conflict parameters, especially conflict intensity and consent from local actors, as its point of departure, it discusses how transnational factors will shape future challenges in international operations. The section defines transnational operations and analyses how transnational conflict challenges the operational requirements for an international military force. The second part presents developments in the legal framework affecting the possibility of obtaining a mandate for international intervention. This part will discuss the mandate parameter, usually a prerequisite for the deployment of an international force. It will shed light on the future possibilities for international and regional consent for intervention. Third, the chapter will explore current operations to assess state-of-the-art mechanisms to address issues on both sides of an international border.

Transnational conflict

A transnational conflict may be defined as a conflict fuelled by cross-border flows. These flows may include people (recruits for armed groups), money (diaspora financing), goods (illegal trade that finances armed groups), ideas (wider ethnic identities or salafist ideology), news (of past atrocities or future popular rallies) and material (weapons). In the future, we expect more transnational conflict, driven by further increased economic interconnectedness and facilitated by further spread and improvements in information technology. These economic and technological trends are a strain on states that exercise a reasonable degree of control over their territory, but become an acute problem when they do not. The increase in transnational conflict is unintendedly strengthening modern international law. In a seminal article, Jackson and Rosberg (1982) said that even the weakest states in Africa would persist because they were recognised by the international community and at the regional level by the African Union, even if they had no control or presence in large parts of their territory, as demanded by the Weberian definition of statehood. The European states emerged before an internationally regulated order was established. In post-colonial Africa, the logic by which the European states were established is turned on its head. All states, no matter how feeble, enjoy the support at least in legal terms of international law and the international community. International law thus guarantees that weak and dysfunctional states are kept alive, partially blocking former colonial areas from developing into stable states (Jackson and Rosberg 1982: 21–22). States with transnational conflict are therefore unlikely to control their borders themselves.

One reason for the relative paucity of thinking on transnational conflicts is the way in which internal conflict used to be understood and studied. Previously, the starting point for most analysis, usually implicit, was that internal conflicts take place within states. This is the 'closed polity approach'. It leads to search for country-specific factors and processes within the state experiencing conflict. This research established that poorer countries experienced more violent conflict than richer ones (Fearon and Laitin 2003). It demonstrated that the relationship between political openness and internal violence is U-shaped; very democratic regimes and very autocratic regimes both experience little violence, whereas intermediate regimes are most prone to experience conflict, especially if they are in the process of moving from one extreme to the other (Huntington 1968; Hegre *et al.* 2001). At the same time, it is a well-established fact that conflicts tend to cluster; some areas are more conflict prone than others. In principle, clustering can be explained within the closed polity approach as a result of shared attributes among the countries in region that make them all more susceptible to civil war.

Within the closed polity approach, it is of course possible to recognise and deal with transnational factors. For example, whether the rebel group has access to an international sanctuary or not is one of the most important factors that set counter-insurgency campaigns apart. If the counter-insurgent manages to control the border, denying the insurgents personnel, weapons and safe havens, the chance of success dramatically increases. Effective border control was present in all eight counter-insurgent successes after 1945, and in only one case did the insurgents win in a case with effective border control (Paul *et al.* 2011, pp. 50–51). Indeed, taking at face value the evidence of all 72 cases after 1945, it appears that effective border control is such a necessary condition for counter-insurgent success that it is quite close to being a sufficient condition. However, good counter-insurgency practices tend to run in packs: counterinsurgents with effective border control engage in other good practices at the same time. Paul *et al.* (2010, p. 51) conclude that 'there is strong evidence in support of border control as a counter-insurgency approach'. More specifically, border control, it seems, may suffice when the insurgency utilises the territory of other states but has only has one target state.

Recent scholarship has begun to analyse transnational conflicts (Buhaug and Gleditsch 2008; Checkel 2013b, pp. 5ff.). Linkages of various kinds between countries explain clustering better than the closed polity approach. Work has begun to uncover how transnational mechanisms cause conflict to spread from one country onto its neighbours (Gleditsch 2007; Cederman *et al.* 2009, pp. 407–408). For legal reasons, the state's authority, although not necessarily its power, drops dramatically at the border. This is why rebel groups, including those not transnational in origin, find it so attractive to operate across a border. These mechanisms are mutually reinforcing and have led to failing states with endemic conflict.

The 'Regional conflict formation perspective' describes how individual and societal objectives may merge into one complex regional conflict (Rubin 2006; Harpviken 2010; Ulriksen 2010). Suggested regionalised conflicts are the 'Hutu-Tutsi nexus' (eastern DRC, Rwanda and Burundi) and the 'Tormented Triangle' (Darfur, eastern Chad and north eastern Central African Republic). Nearly all conditions which create regionalised conflict are present in these examples: politically and ethnically fragmented states that are poor, power vacuums in the hinterland, trans border ethnic groups and a caste of young men whose livelihood and identity depend on their status as combatants (Giroux *et al.* 2010, p. 15). This is a convincing depiction of regional conflict as an entity. The large group of young men is linked to the demographic trends in Chapter 1, and as they depend on being combatants, they are unlikely to consent to settling the conflict.

Thus, the kind of transnational conflicts in which international intervention is most likely share certain characteristics. They will take place in failed or failing states: both the home state of the conflict and the neighbouring host state will be weak states that do not exercise full control over their territory (Bennett 2013, pp. 223–224). They will be grave, with spillover to other areas, creating much loss of life and human suffering. If they were not, the international community would not be sufficiently motivated to engage in them. Transnational operations are highly likely to be operations in which the protection of civilians will be a task (see Chapter 11 in this book).

Consider first how *conflict intensity* will be affected by the conflict being transnational. The economic trends described in Chapter 1 increases the volume of transactions across borders, and the technological trends facilitate communication across borders. The increased transnationality of a conflict contributes to shaping the incentives the combatants face. The altered incentives in turn affect the intensity of the conflict and thus under what circumstances armed groups will be willing to consent to the international force. Trans border kinship ties increases the likelihood of conflict, if the excluded group constitutes a sufficiently large share of the population (Austvoll 2005). Since a diaspora does not bear the full costs of the conflict, it faces incentives that will make it more likely to engage in, to support and to sustain that conflict. Kinship ties can facilitate conflict when they provide resources in terms of money, recruits and military equipment (Adamson 2013, pp. 67, 81). If kinship groups witness or take part in successful insurgency across the border, they become more prone to take up arms themselves (Cederman *et al.* 2009, p. 408). Finally, once conflict breaks out, it may result in decline in trade and investment, which creates economic recession and more refugees, exacerbating the conflict.

Adamson (2013, p. 67) distinguishes between the process of diaspora mobilisation and the impact of diaspora mobilisation on the conflict, once it is mobilised. It is tempting to focus solely on the latter because only mechanisms directly affecting the conflict seem relevant to an international force. However, mechanisms that mobilise the diaspora and the effects of diaspora actions often reinforce each other through feedback loops or overlap in a non-sequential manner (Adamson 2013, p. 68). This is a useful reminder of the complexities a force may meet on the ground.

Countries with many refugees are more likely to experience violent conflict, especially when they come from right over the border. Refugees from faraway conflicts do not increase the probability of conflict in the recipient country but refugees from neighbouring countries experiencing conflict do, and they can be the *cause* of conflict, not only a consequence of conflict. The causal mechanisms are that refugees exacerbate the competition for resources and that they may alter the ethnic balance in the receiving country. Moreover, some of the refugees may be so-called refugee warriors, former soldier with combat experience and an increased tendency to engage in soldierly activities again (Salehyan and Gleditsch 2006, p. 342; Buhaug and Gleditsch 2008, p. 221). Just like ethnic groups and kinship, they facilitate the expansion of social networks that can assist in transfer of arms and other resources (Salehyan and Gleditsch 2006, p. 343). There have always been uprooted soldiers after wars, but, in the future, advances in information technology and increased economic interconnectedness will increase this destabilising potential.

It is natural to think that a settlement allowing refugees to return home signals the end of conflict. Harpviken and Lischer (2013, pp. 90ff.) discuss three potentially destabilising mechanisms rooted in the return of militarised refugees: socialisation to violence, competition for resources and security entrapment. The return of refugees who have been subject to transformative learning fostering militant attitudes is a case that needs extra attention because they are likely to have a lower threshold for engaging in violence. They are, however, similar to all demobilised soldiers domestically displaced or returning from abroad. In the extreme case, these are young men that know nothing but soldiering and whose identity and livelihood depend on their status as combatants. To prevent competition for resources among groups turning violent is a huge task. The struggle among various groups for resources – money, available women to marry, smuggling routes, positions, land, and business opportunities – to attract and sustain membership may lead to conflict. This creates an environment in which warlords thrive.

Warlords are individuals who control small pieces of territory using a combination of force and patronage (Marten 2006, 2012, pp. 187–188). They emerge when state institutions do not function, seeking to control resources through violence. Since warlords make their presence felt in areas with weak or no state institutions, international forces are likely to encounter the phenomenon in some variety. Warlords are not for or against the international force per se. They are on their own side in the competition for resources (Simpson 2012, pp. 44–47). This means that warlords may easily change sides in a conflict. In many cases they will offer an affordable short-term solution to the conflict situation. In Iraq's Anbar province, the Duilami tribe warlords flipped allegiance and turned against the Al-Qaida in Mesopotamia and towards the US military, among other reasons because of competition over smuggling routes, jobs and women. As a consequence, Anbar province went practically overnight from the being the most violent province in Iraq to one of the most peaceful (Long 2008, p. 77).

Warlords are an old phenomenon and it is therefore easy to slip into thinking that they remain the same even in a modern, transnational setting. Unfortunately, modern warlords have limited potential as state builders. Although the Weberian state may have its origins among warlords, warlords are unlikely to be a future source of government because they now face different incentives than they did before international law gave national borders such strong normative status. Globalisation of sovereignty has altered the role of warlords in society. Modern warlords thrive inside states; they defy state sovereignty but they rule through the complicity of state leaders (Marten 2012, p. 3). For example, when Sher Mohammad Akhundzada was removed as governor in Helmand, he became a senator in Kabul. Reduced patronage forced him to disband some of his militia, making fighters available to the Taliban. His militia may even have become a franchise of the Taleban in northern Helmand (Simpson 2012, p. 45). Today, warlords are sheltered from external intervention inside a state and face little pressure to improve governance. Instead, they owe their position to external, transnational patronage with no need to foster internal development through good governance (Marten 2012, p. 190). Moreover, their personal mindset needs to change from being a specialist in violence into being specialists in governance. The fact that warlords rely on personal relationships is not conducive to institution building.

Economic and technological development taking place at the same time made national borders more porous. Warlords as a group will face different incentives in the future because increased interconnectedness give them more options: they may in the future act as middle men in patronage networks that span international borders. Unfortunately, many networks are associated with smuggling of weapons, narcotics, goods (gold and diamonds) and people (Marten 2012, pp. 12–14). The change in warlords to some degree also applies to rebel groups who, unlike warlords, use violence to further political goals but otherwise operate under the same, changed constraints. Rebel leaders, too, are often protected by stronger norms against intervention, and they depend less upon local constituencies for revenue because they are connected to profitable, external networks. Warlords are unlikely to contribute to state building because it would require that they ignore the surrounding incentive structure (Marten 2012, p. 26).

Neighbour states may make bargains with warlords that drain resources or challenge the sovereignty. Warlords can be used to expand influence inside a weak state's neighbour, without actually having to pay the political or material cost of establishing direct control. In 1989, Sher Mohammad's uncle Nasim Akhundzada rose to prominence in Helmand after he issued a fatwa allowing poppy cultivation and after killing 'ten to fifteen khans' (Hafvenstein 2007, p. 129). The same Nasim then offered to ban poppy in Helmand in exchange for American development (Hafvenstein 2007, p. 150). Sher Mohammad Akhundzada did the same thing in 2002, eradicating as much as 85 per cent of poppy in central Helmand, only to abandon the policy when he did not receive the immediate compensation he had expected. Since state weakness is both a cause and a consequence of warlords seeking patronage abroad, the home state is unlikely to be able to regain effective sovereignty on its own. Rwanda has repeatedly been accused of instigating and maintaining the Tutsi M23 group in Eastern DRC to exercise control across the border and weaken the Congolese government.

The relationship between warlords and traditional tribal or ethnic authority structures is complicated. Warlords are not ethnic leaders. In any state likely to experience an international intervention in the future, traditional structures would have long been corrupted. Leaders will therefore operate according to warlord logic and not through legitimate traditional authority structures. However, warlords will belong to a group, and tribal norms and ethnic politics may influence their actions (Giustozzi and Ullah 2006, pp. 9–13; Marten 2012, pp. 188–189). They may also in the future rely on kinship and clanship in their inner circles of advisors and assistants. Sher Mohammad Akhundzada relied heavily on his Alizai tribe for his militia and his family for his network (Hafvenstein 2007, pp. 129–131, 195). Ultimately, what matters is whether a leader owes his position to arms or norms. The trend is towards warlordism and away from traditional authority in areas where an international force is likely to be deployed because both armed conflict and increased interconnectedness erode traditional norms and promote warlordism. An international force may be a valuable source of patronage to a modern warlord. Thus, in the future, international forces will encounter warlords that face incentives that hinder the development of government structures and foster long-term instability. They do not represent a solution to the conflict and relying on them is likely to reduce the prospects of long-term stability.

Let us turn to incentives facing refugees returning to ungoverned areas. They will often find that physical security can only be found as a member of an armed group, often ethnic or tribal in nature. Security entrapment is when the members of a group experience the only answer to an existential threat as being to take up arms. The individual incentives of refugees and the incentives of warlords tend to reinforce one another in ways that exacerbate conflict. In 2001–2005 in Helmand, the Alozai were associated with the government of Afghanistan and the Ishaqzai with the Taliban. For an individual living in Helmand belonging to either tribe, security decided whom to support ('fear'). Individuals, tribes and warlords in Helmand are on their own side. Even though the population may be willing to consent to an international force bringing peace, consent is not secured. Local power brokers profiting from conflict and illegal border transactions are unlikely to consent to measures that destroy their profitable transactions. If the international force is able to contribute to a secure environment promoting economic growth, it may gain the consent of the population, but that is not sufficient if the conflict entrepreneurs do not consent.

In terms of *relative force composition and strength*, at the low end of military capability in transnational operations we find purely criminal gangs. They consist of armed, desperate and violent individuals with a low threshold for violence against anyone threatening their activities. Since an international force will operate among them more than against them, it is necessary for the troops to be prepared for fierce small group combat with a high *individual* experience of conflict intensity. The international force needs to be equipped and prepared to use force in this manner when deployed. Criminal individuals will often attempt to hide their activities by blending in with the civilian population. Therefore the force must avoid collateral damage, while being prepared to fight small groups when it is least expected. The challenges will closely resemble those described by Per M. Norheim-Martinsen in the previous chapter, only in a rural environment.

At the higher end of military capability in transnational operations, we find warlords and militia groups. Even they will have few heavy weapons and limited military capacities. *Conflict intensity* with maximum force applied thus remains relatively rare in transnational conflicts. Operations will typically be at the company, platoon and squad level. If more force is applied, the effect of South African combat helicopters and special forces on M23 is thought provoking. Targeting a militia force like the M23 from the ground and air created an element of jointness that had a powerful effect. The group that had resisted offensives by the FARDC in the past simply fled and disbanded when faced with an airborne weapons system

(Martin 2013). It is both strong evidence of the modest military capacity of such groups, and how effective conventional military power is when faced with armed groups that choose to fight rather than disband. However, the effect is likely to drop sharply as other groups quickly adapt tactics.

In terms of the skill set required, transnational operations might be labelled as counter-insurgency. At the local level in Eastern DRC, however, the economic importance of transnational ties is arguably the driver of conflict. Thus, transnational operation is a more appropriate label, particularly for those armed groups that are not rebelling against anyone. Moreover, the crucial role played by neighbouring states Rwanda and Burundi in instigating and supporting armed groups also make the transnational label highly appropriate. As explained in Chapter 4, how we label operations matters, and in DRC transnational is preferable as it focuses attention on the regional level and the collapse of domestic authority.

Moving on to how transnational conflict shapes the *operational environment* parameter, states with rough terrain are more conflict-prone and likely to experience violent insurgency (Fearon and Laitin 2003). It is also well known that areas with rough terrain within a state experience more violence than others (Buhaug *et al.* 2008). Administrative capacity is one social mechanism explaining why the existence of rough terrain is negatively associated with conflict. The effect of rough terrain is reduced by administrative capacity. Wealthier states therefore experience less conflict as they have better security forces, administrative capabilities and infrastructure. The economic trend of growth moving from Europe and North America to Asia and to sub-Saharan Africa may yield increased resources that may ameliorate the effect of rough terrain in the future.

Rough terrain is also linked to transnational conflict because there is typically more of it in the peripheral areas of a state, i.e. often close to the border. The centre of political authority is typically set in centrally located, welcoming places. This explains the common pattern, at least among relatively weak states, that the state is weaker in peripheral regions (Salehyan 2009, pp. 31–32). Typically, the government first loses control in the mountains of Afghanistan, the deserts of Mali, the jungles of DRC and the swamps of Sudan. Thus the troops in transnational operations nearly always face the challenge of controlling long lines of communication in difficult terrain. Moreover, the mission often should ideally have the capability to dynamically monitor networks in rough terrain.

In transnational operations, two capabilities are particularly important: logistics and intelligence. Logistical capability vehicles and helicopters are not only force multipliers but also a necessary condition to achieve utility of force. Transnational operations are intelligence intensive. The intelligence activities and demands are not necessarily different from those noted in Chapter 2 on high intensity warfare, in Chapter 3 on counter-insurgency operations and in Chapter 11 on protection of civilians. They are, however, compounded by two additional needs. First, it is necessary to collect and analyse information from two or more states. Second, they will often need to penetrate closed networks operating on both sides of an international border.

Finally, transnational mechanisms do not have a deterministic relationship with conflict. They will certainly lead to challenges, but, when these are adequately addressed, conflict can be avoided. The fighting in neighbouring Mozambique during the 1980s and early 1990s led nearly two million people to seek refuge in Malawi, one of the poorest countries in the world. During the worst phase, refugees constituted 10 per cent of Malawi's resident population, setting many of the abovementioned mechanisms in motion. However, in spite of limited resources, Malawi provided a strong local response, with considerable help from the United Nations High Commissioner for Refugees (UNHCR), the World Food Program and a range of NGOs (Salehyan and Gleditsch 2006, p. 361). The point is that a strong local and international response can mitigate and indeed prevent the spread of conflict, even though limited resources are typically negatively associated with an effective policy response. An international force may also assist in containing or preventing conflict.

Transnational mandates

Turning to the conditions for establishing an international intervention, this section explores the conditions for obtaining a transnational mandate. One obvious response to transnational conflict is to have fully fledged transnational operations. In 2007, the UN in Operation Artemis was mandated for the first time to operate in both neighbouring Central African Republic and Chad, in response to the on-going humanitarian refugee crisis in Sudan's Darfur province. In May 2007, the French government requested the EU member states to do 'something in Eastern Chad' (Norheim-Martinsen 2011, p. 24) to protect the many refugees in Chad and the Central African Republic. It was the first UN operation with a regional mandate in response to a regional problem. The problems in the tormented triangle are certainly transnational but Operation Artemis was a limited, short-term operation. In Chapter 11, Alexander W. Beadle will describe how sovereignty has been modified from simply a right to non-interference to also include an obligation to protect one's own population from mass atrocities. Protection of civilians is not necessarily a concern or the reason for a mandate to operate in two or more states, but it could be.

An operation also needs consent from regional and international actors. While it is harder to make two states consent than one, the problem may be compounded if the two states have an antagonistic relationship, or opposite interests in the conflict that makes international intervention desirable in the first place. For example, Pakistan has refused to make its territory available to create a genuine AfPak ground strategy. Another way is for intervention in a deeply troubled area to be undertaken in the name of the international community. The remainder of this section will discuss the possibilities and the many obstacles for future fully fledged transnational operations, taking future trends in international and regional consent as its starting point.

Increased interconnectedness yields increased respect for the norm of international sovereignty (Zacher 2001). The norm emerged in the aftermath of World War I; it became accepted after World War II as a cornerstone for the UN Charter 2(4), and it has become institutionalised and increasingly observed since. International regional organisations in Africa and the Americas have been established, based on the principle of territorial respect (Zacher 2001, p. 234). As noted in the section above, the economic trends described in Chapter 1 have given weak states more resources at their disposal, increasing their capacity to enforce a consistent set of rules over their entire territory. Territorially based state power is more needed than ever to enforce the rules of an international economy (Zacher 2001, pp. 245–246). For example, the need for a functioning state to guarantee investment and property rights may arguably be the driver behind some of China's support for African governments. From another source, positive pressure from democracy in support of the sovereignty norms has been suggested. Democratic regimes value territorial integrity higher than ethnic self-determination because it best serves their values of liberty and order while eliminating competing political principles (Zacher 2001, p. 245). However, even though economic and political development will continue to remove the causes for international intervention in some areas, plenty of failed and failing states remain. Negative feedback from the lack of economic and social development can fuel conflict to such an extent that it seems unlikely that order can be restored without foreign assistance. The strengthening of the international sovereignty norms does mean, however, that a mandate to operate on both sides of a border is a necessary condition for such an intervention to take place.

Who would authorise such a mandate and under what circumstances? Regional security organisations may have a useful role to play in transnational operations. The emergence of no fewer than 38 regional organisations with a security mandate reflects the urgency of the issue (Tavares 2009, p. 2). The UN has also recognised the challenges of regional conflicts from the beginning. It uses the options presented in the UN Charter. Chapter VIII Regional Arrangements paragraph 52 gives regional organisations a role in settling local conflicts. In paragraph 53, the Security Council is encouraged to utilise 'regional arrangements or agencies for enforcement action under its authority'.

The consent of regional organisations and neighbouring states may be helpful for the international force. Regional actors and institutions know the neighbourhood and may share culture, languages and local politics. Local actors will be part of local politics and may exercise influence through personal relationships or control of resources that may help the intervention.

Since they are the ones most affected by the conflict, they have a legitimate interest in containing it (Tavares 2008, p. 111). At a minimum, regional actors that are stakeholders in the intervention will not thwart or sabotage the operation. From an international perspective, regional institutions provide legitimacy. Intervention will not so easily be portrayed as Western self-serving meddling when it is a response to a regional request to deploy military forces. For example, the Economic Community of Western African States (ECOWAS) deployed in the wake of the French intervention in Mali in 2012, at least partly in order to play this role.

Moreover, regional organisations and states may give the necessary legitimacy to obtain an international mandate in the UNSC. Let us look at some of the powers that have had recent economic success, to reflect on international consent in the future. China has a restrictive view of intervention, supporting the norm of territorial respect (Zacher 2001, p. 233; Gill and Huang 2013, p. 140), but also tends to judge important issues on a case-by-case basis (Gill and Huang 2013, p. 152), taking into account possible state collapse, no effective actor whose sovereignty can be violated, humanitarian concerns and regional support. Russia might have an even more restrictive view of international intervention, and tends to lump interventions, be they UN peacekeeping, UN mandated operations or operations without a mandate, together in the same basket (Nikitin 2013, pp. 163, 174–175). Russia also tends to view all such interventions as a tool for great power influence. On the other hand, Russia has a strong preference for regional instruments to manage intervention in a crisis, particularly, it must be said, in its own neighbourhood through institutions dominated by Russia (Nikitin 2013, pp. 164–168, 177). In March 2013, regional support swayed both China and Russia to vote in favour of UNSC Resolution 2098, leading to a unanimous authorisation of the Intervention Brigade in DRC. To sum up so far, transnational operations are unlikely in functioning states, leaving only failing or failed states with large ungoverned areas. Regional security organisations provide necessary legitimacy, both in the region and in the UNSC, to achieve consent to mandate the operation. Moreover, regional organisations may provide the frame for the military operation, contributing invaluable local knowledge, relevant leverage and the necessary military forces for the operations.

Mandates and operations today, however, are not fully-fledged transnational operations across an international border. There is a need to explore other ways in which the operations can have the intended effects across an international border.

Transnational mechanisms

Let us turn to some examples of how contemporary transnational operations have handled the growing awareness of the need to impact both sides of an international border. Since 1999, a UN mission in DRC has been grappling with transnational challenges from the Tutsi–Hutu nexus in Eastern Congo, Rwanda, the Central African Republic and Burundi. It has been a frustrating experience but one that has given time to learn and improve. Many cutting-edge concepts have evolved bottom-up in an attempt to solve concrete challenges (Kjeksrud and Ravndal 2010, especially p. 40; Kjeksrud and Ravndal 2011). Similarly, decisions at the political–strategic level may be shaped by concrete challenges, and the decision to create an Intervention Brigade in MONUSCO is an example. The purpose of the following section is thus to examine how transnational challenges are being dealt with in operation MONUSCO, without pretending to provide a comprehensive analysis of the conflict in DRC.

The first mechanism for handling transnational conflict is a regionally backed international combat force. In September 2012 at the International Conference on the Great Lakes Region (ICGLR), leaders in the African Union and United Nations decided to immediately establish a 4000-strong neutral international force to bring to the region. In March 2013, the Security Council adopted Resolution 2098, authorising an Intervention Brigade and extending the mission's mandate until 31 March 2014. The resolution represents state-of-the-art thinking about what to do about transnational dynamics in international operations. To begin with, the resolution acknowledged 'that eastern DRC has continued to suffer from recurring cycles of conflict and persistent violence by armed groups, both Congolese and foreign' (United Nations 2013). The resolution was a response to a regional initiative, the so-called Peace, Security and Cooperation Framework for the Democratic Republic of the Congo and the region ('the PSC Framework'). The PSC Framework consisted of the AU Commission, the South African Development Community and ICGLR, as well as the 11 most relevant neighbouring countries, including Rwanda, Burundi and Uganda. The framework is certainly comprehensively regional, but has so far lacked a concrete plan to achieve its lofty ambitions. It has, however, served the purpose of legitimising the Intervention Brigade.

The state failure in Eastern DRC has resulted in an array of armed groups and violent fighting among them and with the Congolese army FARDC. Militarily these groups, including the Tutsi M23 and its main foe, the almost entirely Hutu-based FDLR (*Forces démocratiques de libération du Rwanda*), were not remarkable. They have also significantly been losing strength since 2001, although information is unreliable. In early 2013, the remains of M23 were believed to be about 1,500 troops and the FDLR somewhat more. The M23 was the latest of a number of Tutsi groups backed by Rwanda's Tutsi government to fight against the Burundi-backed Hutu FDLR. On 18 March 2013, M23 leader Bosco Ntaganda turned himself in at the US embassy and requested transfer to the International Criminal Court. A splintering within the M23 movement, with Ntaganda's forces suffering a major defeat after only two days, and the withdrawal of support from his sponsor Rwanda figure featured most prominently among the explanations for his surrender (Reuters 2013). Numerous previous efforts have been made by the Congolese forces to dislodge the various Tutsi groups operating in Eastern Congo, with or without support from MONUSCO. On 20 January 2009, the Rwandan Army, in concert with the Congolese government, even entered the DRC to hunt down FDLR fighters. But the efforts to degrade the military capacity of the M23 and FDLR and their predecessors have consistently failed to remove them permanently because the rebels just run and hide, often on the other side of the border.

The following transnational devices to solve the rebel problem in DRC are possible, in addition to MONUSCO's support for national Congolese efforts. The first is outright intervention to support Congolese efforts, through the state sponsoring one rebel group. In 2009, Rwanda government troops entered the DRC to help in the fight against FDLR. It failed then and is likely to fail again because it leads to increased support of the FDLR from Burundi. The second is some kind of genuinely regional force to defeat the rebels. This is fraught with political problems due to low level of trust in the region and the Hutu–Tutsi security dilemma. The third is troops and political support from the main regional states outside the conflict (Tanzania, Malawi, Mozambique and South Africa), under the auspices of the UN.

The Intervention Brigade, which deployed in mid-2013, falls within the last category. Its deployment was supplemented with political pressure from a united international society on Rwanda, combined with reassurances that the threat from the Hutu FDLR would also be addressed. The Intervention Brigade deployed in mid-2013. In conjunction with the political process that made it possible, it was also capable of overpowering any rebel group operating in Eastern Congo, making it potentially the best attempt yet to come to terms with the bewildering transnational conflict dynamics in Eastern Congo. The International Brigade aided a FARDC offensive against the M23 that secured the first military victory of Congolese forces over a violent rebel group. This surprising turnaround is mainly attributed to four factors, three of which have a decidedly transnational dimension. First, the fighting capabilities of the FARDC had improved. Second, due to international pressure to withdraw aid, Rwanda did not back the M23 this time around. Third, the South African combat helicopters and special forces in the Intervention Brigade provided firepower that the M23 could not match (U.S. Army Peace and Stabilization Operations Institute 2013; Martin 2013; Stearns 2013). Fourth, the Dodd–Frank transparency bill passed by the US Congress had undermined funding for rebel groups in DRC. The legislation had made regional governments, multinational companies and local civil society organisations undertake series of actions that had removed much of the profitability of rebellion by replacing national control of transnational flows with international transparency. Gold is the exception (Prendergast 2013).

In conclusion, the transnational effort that made the Intervention Brigade possible seems to have done what military power can do: it has created the conditions for a political solution (Smith 2006). Whether that political solution is forthcoming remains to be seen. The main political obstacle is clearly to find a way that alleviates the existential threat that Rwanda's Tutsi government experiences from the Hutu FDLR. Reportedly, the remains of the M23 rebels still occupy the hills along the Rwandan border between Runyoni and Tshanzu (Stearns 2013).

A second transnational mechanism is to have national forces operating in each country. An example is provided by the latest attempt to eliminate the Lord's Resistance Army (LRA). LRA originated in Aioch province in Uganda but when pressured has moved into neighbouring countries. Now DRC, South Sudan and Uganda attempt to coordinate their national forces operating within each home state, share intelligence and have also granted *a mutual right to hot pursuit,* to be more effective in hunting down the LRA's soldiers. This example of international cooperation has been facilitated by the fact that the LRA has no transnational support and is a common menace to all.

A third mechanism is the transnational repatriation between DRC and Rwanda. Dealing with former soldiers is usually referred to by the acronym DDR, i.e. disarmament, demobilisation, and reintegration. In operation MONUSCO, DDR has two operational components. The first is conventional DDR, focusing on domestic Congolese armed groups. The second component has acquired two additional 'Rs' in its acronym (DDRRR), denoting resettlement and repatriation to better deal with the array of illegal foreign armed groups in the DRC. These include the FDLR, Allied Democratic Forces, LRA, and National Front for Liberation – Burundi. The objective is to repatriate all foreign fighters to their country of origin.

A final mechanism set up to meet the additional transnational demands of the DDRRR process is so-called sensitisation directed towards foreign fighters. The idea is inform fighters of what they need to do to enter the DDRRR process. They are brought up to date on unfolding developments in their home countries, and they are informed about the packages and benefits they will accrue if they voluntarily enter (MONUSCO 2013). MONUSCO has an operative concept to deal with fighters who do want to lay down their arms and return to civilian life in their home country. MONUSCO forces provide secure passage to transit centres inside DRC, where the fighters stay for four days at the most. The transit centres give them access to communication so they can contact relatives at home and provide bus transportation out of Congo. Civilians are resettled by the UNHCR, whereas ex-combatants stay for 45 days preparing for civilian life, in centres run by the Rwanda Demobilization and Reintegration Commission. The Ugandan Amnesty Commission, and the South Sudan Commission for DDR provide similar services for ex-combatants in DRC from these countries. This transnational system was in place and was, for example, called upon by the UNSC to demobilise the hundreds of M23 combatants who followed General Ntaganda in his flight from the DRC into Rwanda on 18 March 2013. This transnational mechanism provides an exit strategy for fighters that enable them to leave their armed group and gives them safe transport out of DRC, ameliorating conflict. In addition, if successful, it avoids leaving the country of origin with a destabilising ex-fighter group.

Conclusion

The trend in Chapter 1 unambiguously suggests transnational conflict is with us and on the rise. Transnational operations involve an extraordinarily complex use of force. The military tasks themselves will be similar to normal military activities, but there is a need to develop a strategy to harmonise military and political approaches and to balance the short-term need to end fighting with long-term institutional reform. Again, these kinds of challenge are in principle known in stabilisation operations, but the number of actors and the additional political and legal issues created by international boundaries add extra layers of complexity to the strategy of transnational operations. Indeed, the international ramifications are so complicated that the efforts to find a mandate or consent to a transnational operation are often thwarted. That should not distract from the fact that creating transnational impact is the best remedy for the situation, and that it may be possible to influence without a mandate.

Indeed, the realistic answer is not necessarily a fully-fledged transnational operation with full transnational mandate. There is a need to find a way to make the international response more transnational to effectively address transnational conflict even where a mandate to operate in two states is difficult to obtain. The creative responses to the challenges in DRC are of general interest in this respect. The UN and the international community are slowly improving their skills and arsenal of techniques to cope with transnational challenges. Regional backing of an intervention force, transnational repatriation of former combatants or mutual agreements between states that allow them to operate forces on the territory of one another, are ways to make the international more transnational. The MONUSCO operation in DRC reveals that there is an array of possible ways to influence across an international border without a direct mandate. To develop such transnational mechanisms further seems to offer the most realistic and useful way to address future transnational challenges.

References

Adamson, Fiona B. (2013). "Mechanisms of Diaspora Mobilization and the Transnationalization of Civil War". In Jeffrey T. Checkel, Ed. *Transnational Dynamics of Civil War*. Cambridge: Cambridge University Press.

Austvoll, Martin "Transnational Ethnic Dimensions of Third-Party Interventions in Civil Conflicts". Master's thesis, University of Oslo.

Bennett, Andrew (2013). "Causal Mechanisms and Typological Theory in the Study of Civil Conflict". In Jeffrey T. Checkel, Ed. *Transnational Dynamics of Civil War*. Cambridge: Cambridge University Press.

Buhaug, Halvard and Kristian Skrede Gleditsch (2008). "Contagion or Confusion? Why Conflicts Cluster in Space". *International Studies Quarterly* 52(2), 215–233.

Buhaug, Halvard, Lars-Erik Cederman and Jan Ketil Rød (2008). "Disaggregating Ethno-nationalist Civil Wars: A Dyadic Test of Exclusion Theory". *International Organization* 62(3), 531–551.

Cederman, Lars-Erik, Luc Girardin and Kristian Skrede Gleditsch (2009). "Ethnonationalist Triads: Assessing the Influence of Kin Groups on Civil Wars". *World Politics* 61(03) 403–437.

Checkel, Jeffrey T. (2013a). *Transnational Dynamics of Civil War*. Cambridge: Cambridge University Press.

Checkel, Jeffrey T. (2013b), "Transnational Dynamics of Civil War". In Jeffrey T. Checkel, Ed. *Transnational Dynamics of Civil War*. Cambridge: Cambridge University Press.

Fearon, James D. and David D. Laitin (2003). "Ethnicity, Insurgency and Civil War". *American Political Science Review* 97(1), 75–90.

Gill, Bates and Chin-Hao Huang (2013). "The People's Republic of China". In Alex J. Bellamy and Paul D. Williams, Eds. *Providing the Peacekeepers: The Politics, Challenges and Future of United Nations Peacekeeping Contributions*. Oxford: Oxford University Press.

Giroux, Jennifer, David Lanz and Damiano Sguaitamatti (2010). "The Tormented Triangle: The Regionalisation of Conflict in Sudan, Chad and the Central African Republic". Crisis States Research Centre working papers series 2, 47. Crisis States Research Centre, London School of Economics and Political Science, London.

Giustozzi, Antonio and Noor Ullah (2006). *'Tribes' and warlords in southern Afghanistan, 1980–2005*. London: Crisis States Research Centre, London School of Economics and Political Science.

Gleditsch, Kristian Skrede (2007). "Transnational Dimensions of Civil War". *Journal of Peace Research* 44(3), 293–309.

Hafvenstein, Joel (2007). *Opium Season: A Year on the Afghan Frontier*. Guilford Conn: The Lyons Press.

Harpviken, K. B. 2010. "Troubled Regions and Failing States: Introduction". In Harpviken, K. B., Ed. *Troubled Regions and Failing States: The Clustering and Contagion of Armed Conflicts*. Bingley, UK: Emerald.

Harpviken, Kristian Berg and Sarah Kenyon Lischer (2013). "Refugee Militancy in Exile and upon return to Afghanistan and Rwanda". In Jeffrey T. Checkel, Ed. *Transnational Dynamics of Civil War*. Cambridge: Cambridge University Press.

Hegre, Håvard, Tanja Ellingsen, Scott Gates and Nils Petter Gleditsch (2001). "Toward a Democratic Civil Peace? Democracy, Political Change, and Civil War, 1816–1992". *American Political Science Review* 95(1), 33–48.

Huntington, Samuel, P. (1968). *Political Order in Changing Societies*. New Haven, Conn: Yale University Press.

Jackson, Robert H. and Carl G. Rosberg (1982). "Why Africa's Weak States Persist: The Empirical and the Juridical in Statehood". *World Politics* 35(1) 1–24.

Kjeksrud, Stian and Jacob A. Ravndal (2011). "Emerging Lessons from the United Nations Mission in the Democratic Republic of Congo: Military Contributions to the Protection of Civilians". *African Security Review* 20(2), 3–16.

Kjeksrud, Stian and Jacob Aasland Ravndal (2010). "*Protection of Civilians in Practice – Emerging Lessons from the UN Mission in the DR Congo*". Kjeller: Norwegian Defence Research Establishment.

Long, Austin (2008). "The Anbar Awakening". *Survival* 50, 67–94.

Marten, Kimberly (2006). "Warlordism in Comparative Perspective". *International Security* 31(3), 41–73.

Marten, Kimberly (2012). *Warlords: Strong-Arm Brokers in Weak States.* Ithaca and London: Cornell University Press.

Martin, Guy (2013). "DRC Sniper Revelation Compromising SANDF Troops". *Defence Web*, 5 September.

MONUSCO (2014). *Sensitations Mechanisms for Foreign Armed Groups.* www.monusco. unmissions.org/Default.aspx?tabid=10725&language=en-US

Nikitin, Alexander (2013). "The Russian Federation". In Alex J. Bellamy and Paul D. Williams, Eds. *Providing Peacekeepers: The Politics, Challenges and Future of Un Peacekeeping.* Oxford: Oxford University Press.

Norheim-Martinsen, Per Martin (2011). "Our work here is done: European Union Peacekeeping in Africa". *African Security Review* 20(2) 17–28.

Paul, Christopher, Colin P. Clarke and Beth Grill (2010), *Victory has a thousand fathers: Sources of Success in Counterinsurgency.* Santa Monica, Calif: Rand Corporation.

Paul, Christopher, Colin P. Clarke and Beth Grill (2011). *Victory Has a Thousand Fathers: Evidence of Effective Approaches to Counterinsurgency, 1978–2008.* Small Wars Foundation, 13 January.

Prendergast, John (2013). "Daily Beast Op-ed: How Congo Defeated the M23 Rebels". In *Daily Beast: Enough! The project to end genocide and crimes against humanity.* www.enoughproject.org

Rubin, B. R. (2006). "Central Asia and Central Africa: Transnational Wars and Ethnic Conflicts". *Journal of Human Development* 7, 5–22.

Salehyan, Idean (2009). *Rebels Without Borders: Transnational Insurgencies in World Politics.* Ithaca, NY: Cornell University Press.

Salehyan, Idean and Kristian Skrede Gleditsch (2006). "Refugees and the Spread of Civil War". *International Organization* 60(2), 335.

Salehyan, Idean, Kristian Skrede Gleditsch and David E. Cunningham (2011). "Explaining External Support for Insurgent Groups". *International Organization* 65(4), 709–744.

Simpson, Emile (2012). *War from the Ground up: The twenty-first-century Combat as Politics.* London: Hurst.

Smith, R. (2006). *The Utility of Force: The Art of War in the Modern World.* London: Penguin Books.

Stearns, Jason (2013). "As the M23 Nears Defeat, More Questions than Answers". In Jason Stearns, Ed. *Congo Siasa.* http://congosiasa.blogspot.co.uk/ 2013.

Tavares, Rodrigo (2008). "Understanding Regional Peace and Security: A Framework for Analysis 1" *Contemporary Politics* 14(2) 107–127.

Tavares, Rodrigo (2009). *Regional Security: The Capacity of International Organizations.* London and New York: Routledge.

Ulriksen, S. (2010). "Webs of War: Managing Regional Conflict Formation in West and Central Africa". In K. B. Harpviken, Ed. *Troubled Regions and Failing States: The Clustering and Contagion of Armed Conflicts*. Bingley, UK: Emerald.

United Nations. (2013). S/RES/2098 www.un.org/en/ga/search/view_doc.asp?symbol=S/RES/2098(2013)

US Army Peacekeeping and Stability Operations Institute. "UN Force Intervention Brigade against the M23". In SOLLIMS – Stability Operations Lessons Learned and Information Management System, 2013.

Zacher, Mark W. (2001). "The Territorial Integrity Norm: International Boundaries and the Use of Force". *International Organization* 55(2), 215–250.

10 Cyber operations

Siw Tynes Johnsen

'Cyberwar is coming!' announced John Arquilla and David Ronfeldt as early as 1993, as they claimed that the emergence of cyberspace would transform the very nature of warfare. Today, there is no shortage of prominent figures claiming that cyber war has arrived. Richard Clarke, former Special Advisor to the President for Cyber Security in the White House, claimed in 2010 that 'cyberwar has begun', and in February 2011 then-CIA Director Leon Panetta warned the US House Permanent Select Committee on Intelligence that 'the next Pearl Harbor could very well be a cyber attack'. After the highly sophisticated computer worm Stuxnet damaged the Iranian nuclear programme, a much-noted investigative article in Vanity Fair concluded that the event foreshadowed the destructive new face of twenty-first century warfare, saying that 'Stuxnet is the Hiroshima of cyber-war' (Clarke 2010).

Rapid technological developments have enabled armed forces around the globe to take advantage of cyberspace to achieve effects in military operations. While the cyber domain has to a large extent been recognised as a new domain of warfare, alongside land, sea, air and space, this chapter will argue that it is doubtful whether future military operations will take place exclusively within the cyber domain. Understanding the nature of the cyber domain makes it clear that it will not change the nature of warfare. Future bloodless wars fought online are considered more science fiction than reality – wars will still be fought 'amongst the people' (Smith 2006). However, it is safe to expect that any future war will have a cyber component. Such components range from espionage via information operations to attacks seeking to cause direct damage to systems on which the enemy is dependent.

This chapter will begin by tracing the evolution of the cyber domain and its role in military operations. It will then assess the current situation surrounding the use of the cyber domain in armed conflict. Finally, the chapter will explore how this may change, looking at how the cyber domain may feature in future military operations. This will be done by examining the cyber domain's effect on the five parameters sketched out in Chapter 2 – *mandate, consent, operational environment, relative force composition and strength*; and *conflict intensity* – and what particular challenges forces may face in this respect.

Evolution of the cyber domain and military operations

The cyber domain is defined as the physical and logical interconnection of information systems, including network devices, communications infrastructure, media and data. For the purpose of this chapter, the terms 'cyber domain' and 'cyberspace' will be used interchangeably. In the cyber domain, actors wield cyber power: the ability to apply or project power in or through the cyber domain. This can be done, for example, through conducting military operations in or through the cyber domain, so-called cyber operations (Windvik and Diesen 2013).

Societies are growing increasingly dependent on the cyber domain and the armed forces are no exception. During the first Gulf War, the United States and coalition forces used information and communications technology to their advantage, severing Iraqi lines of communications and disrupting Iraqi command and control. Due to these cyber efforts, the war is termed 'the first information war' (Campen 1992). While this type of warfare was not known as cyber warfare at the time, Arquilla and Ronfeldt described US application of 'cyberwar principles' as 'superior' compared to those of their adversary. They claim Iraq's 'organization, doctrine, strategy and tactics were from a different era' (Arquilla *et al.* 1993, p. 38).

While the cyber means used by coalition forces during the first Gulf War were relatively advanced at the time, developments have since taken place at an extremely rapid pace. The first United States joint offence–defence cyber command was established in 2000, the Joint Task Force – Computer Network Operations (JTF-CNO). Its first commander, Major General James D. Bryan, stated that at this point they 'were still in [their] infancy in terms of fully appreciating the effects we could achieve' when considering cyber operations (Healey 2013, p. 58). Jason Healey argues that, after 2003, the militarisation of cyber conflict increased, and the offensive capabilities within JTF-CNO were transferred to the National Security Agency, under operational control of US Strategic Command. Major General Bryan noted that, with offensive cyber operations, 'you sometimes see the failures. But you rarely see the successes' (Healey 2013, p. 67). Thus, the most successful cyber operations are perhaps still unknown to the public.

Cyber operations seem especially well suited early in a conflict, as a way of applying political pressure without resorting to physical attacks and infringement of another state's territory (Diesen 2013). An example of this took place in 2007, after Estonian authorities decided to move the Bronze Soldier War Memorial, which was located next to the military cemetery in Tallinn. This decision was instantly characterised by Moscow as a defamation of the Soviet dead and the fallen soldiers buried nearby.

Tensions grew in Estonia as riots between a Russian ethnic group and an Estonian nationalist group broke out on 27 April the same year (Clarke 2010). The fact that these tensions quickly entered the cyber domain did not come as a surprise to Estonians. The Estonian cyber security community assumed that 'when there are riots in the streets, they will eventually go cyber' (Schmidt in Healey 2013, p. 178). Estonian government networks were harassed by distributed denial of service (DDOS) attacks of a magnitude described by Clarke as 'the largest ever seen', which disrupted Estonians' access to online banking, newspapers and a range of government services (Clarke 2010; Schmidt in Healey 2013). The United States Computer Emergency Response Team (US CERT) describes DDOS attacks as an attempt to prevent legitimate users from accessing information or services, usually by 'flooding' a target with requests. As a system can only process a certain number of requests at once, such floods render the server unable to provide the services intended (US CERT 2013). Estonia claimed that they traced the attacks back to computers located in Russia; however, the Russian government denied any involvement in the cyber attacks and refused to assist Estonia in identifying the computers on Russian territory. This illustrates how difficult attribution is in the cyber domain. Some Russian government officials admitted, however, that the attacks could have been the result of 'patriotic hackers', civilians taking matters into their own hands. It is arguably more common for former communist states to use civilian hackers as part of the state apparatus (Clarke 2010; Schmidt in Healey 2013; NATO Review 2013; Kramer 2011). These cyber attacks were raised to the level of the North Atlantic Council and led to the 2008 establishment of the Cooperative Cyber Defence Centre of Excellence in Estonia's capital, Tallinn. While this incident had limited impact on military operations per se, it certainly impacted NATO with regard to protecting against and responding to attacks in cyberspace.

While the cyber attacks on Estonia were not part of a larger military offensive, the following year showed efforts in cyberspace which were probably coordinated with actions of conventional military forces. This is probably the most effective way of using cyber means to reach military objectives, by conducting cyber operations in concert with operations in the traditional military domains. This can be done not only as a show of force but as a force multiplier in order to take down services at a critical point in an operation, reducing an adversary's situational awareness (Diesen 2013). The conflict between Georgia and Russia erupted as Russian populations in the Georgian territories of South Ossetia and Abkhazia defeated the Georgian army and expelled most Georgians. Russian authorities supported the secessions and funded 'independent' governments set up to rule the territories. South Ossetian rebels staged missile attacks on Georgian villages, which led to military confrontation between Georgia and Russia. At the exact same time as the Russian army moved, Georgian government websites and online news media were targeted in a

DDOS attack. This was at the time seen as a rather innovative move, and the intention seemed to be to make sure that Georgia would not know when and how the Russian forces were moving, degrading their situational awareness. As it did with regard to the attacks in Estonia, the Russian government claimed that the cyber attacks were conducted by private individuals not connected with national authorities (Clarke 2010; NATO Review 2013).

The current situation

While conventional military means have gained more attention than cyber means during the war in Afghanistan, the cyber domain has been utilised in this conflict as well. Lt General Richard P. Mills in the United States Marine Corps said, 'I can tell you that as a commander in Afghanistan in the year 2010, I was able to use my cyber operations against my adversary with great impact. I was able to get inside his nets, infect his command-and-control and in fact defend myself against his almost constant incursions to get inside my wire, to affect my operations' (Healey 2013, p. 66; National Public Radio 2013). His statement was described by National Public Radio as a 'rare acknowledgment that the military engages in offensive cyber operations' (National Public Radio 2013). The American forces were, however, not alone in being targeted in the cyber domain. Among few publicly known cases, Norwegian and Danish closed military networks were penetrated in 2008 and 2009 respectively. The Head of the Norwegian Cyber Defence College, Lt Colonel Roger Johnsen, told Norwegian broadcaster NRK that the likely perpetrator was foreign intelligence seeking to gather information and if possible degrade the systems in a critical situation. Danish forces acknowledged that there had been an 'IT incident', but declined to comment on the incident any further (DR online 2012; NRK online 2012).

In April 2012, Taliban spokesman Zabihullah Mujahid told Reuters that the Taliban website had just been hacked for the third time in less than a year, blaming Western intelligence agencies. The original text on the Taliban website had been replaced with the message 'Violence is wrong in all its forms, especially the encouragement by the Taliban of cowardly betrayal and the senseless murder of innocent civilians'. This type of cyber attack is called defacement, where a website's original content is replaced with that of the attacker (Reuters online 2012). The previous year, the al Qaeda online publication Inspire Magazine fell victim to this same tactic, as one of its recipes for 'how to make a bomb in the kitchen of your mom' was replaced with recipes from the Ellen DeGeneres Show featuring the 'best cupcakes in America'. This act of defacement is known as Operation Cupcake and was conducted by British intelligence agency MI6. After the fact, it became known that US intelligence agencies had also had access to the magazine, but chose not to reveal the fact (Washington Post online

2011a). These two intelligence organisations came down on different sides of what has become a classic dilemma in cyberspace: once you have gained access to a system or website, you can choose to use it for intelligence gathering, staying hidden, or you can use that access to alter information, knowing that by doing so you are losing the intelligence source.

Stuxnet, which was discovered in 2010, is perhaps the best known 'cyber weapon' seen to date. It was a complex piece of malware designed to interfere with the Siemens industrial control systems used by Iran in their nuclear enrichment centrifuges in Natanz. While never confirmed, it was allegedly created by the United States and Israel as part of operation Olympic Games, aimed at damaging the Iranian nuclear programme. Designing malware of this size and sophistication would have required the resources of a large and technologically advanced nation-state. The malware made the nuclear centrifuges spin faster and faster out of control, which finally caused them to break down. This is the only publicly acknowledged cyber attack known to have caused physical damage, which was only possible because the centrifuges were designed in a way which would allow them to spin faster than the structure could handle in the first place. This would be extremely difficult, if not close to impossible, to replicate. However, developments in the cyber domain take place at an extremely rapid pace, which means we cannot exclude this completely from future scenarios (Kilcullen 2013, pp. 177–178; Morton in Healey 2013).

Towards the future in the cyber domain

Through media coverage, one may get the impression that any technologically savvy individual with a computer can incapacitate large nation-states without much difficulty by utilising the cyber domain. However, this ignores the fact that it is very difficult to conduct complex attacks with serious consequences in the cyber domain. While taking control of a random computer might be easy, it is very demanding to take control of a specific target. This requires, for instance, intimate and detailed knowledge of operating systems and specific configurations, which generally requires extended intelligence efforts in advance. Because of this, it is much easier for actors committing cyber crime, as they can go for the low-hanging fruit: using botnets to get control over easily accessible computers with known vulnerabilities for exploitation (Windvik and Diesen 2013).

Using cyber means as part of advanced military operations, however, is even more demanding in terms of finances, required skills and time. In many cases one would need to build an exact copy of the adversary's system with all of the exact same configurations to perform extensive testing before executing a live attack with a cyber weapon. According to the *New York Times*, the United States decided against using offensive cyber operations against Libyan air defence systems in 2011, in part because of concerns about 'whether the attack could be mounted on such short notice' (*New York Times* online 2011; Healey 2013).

These weapons are to a large extent 'one shot guns' – they are released into the adversary's networks, and once they have been used, the adversary will be able to patch vulnerabilities and make sure that this exact weapon will not be used successfully against them again. Using a cyber weapon is therefore largely the same as sharing the cyber weapon with the world (Windvik and Diesen 2013). An analogy from the air power domain is to design and construct a superior type of fighter jet, penetrate the adversary's anti-air defences and attack your adversary but then land the jet on the adversary's air field, leave the aircraft and letting the adversary spend all the time in the world inspecting and studying the plane afterwards. Your advantage has effectively been terminated, and in a way the playing field has yet again been levelled. To be successful in using cyber means as part of an advanced military operation, one would also need a large and highly developed intelligence organisation (Diesen 2013).

Due to the high complexity and demand for resources involved in conducting serious attacks in the cyber domain, future bloodless wars taking place in the cyber domain alone are still more science fiction than reality. Thus cyber warfare does not represent a new and separate form of warfare but is a complementary means to other types of force and weapons systems. When used with utility, cyber efforts will serve as a powerful force multiplier. Modern militaries are gradually realising this and are making efforts to take advantage of this new warfighting domain (Diesen 2013).

Mandate

This section will explore challenges related to the cyber domain and both the *jus in bello* and *jus ad bellum* of armed conflicts today. While the topics have been fiercely debated, most legal scholars have come to the conclusion that cyber warfare does not represent such a change in warfare that it falls outside being regulated by already existing international law. In fact, an international consensus seems to be forming around the position that existing international law regimes also cover cyber means. The law just needs to be interpreted in a way that encompasses the specificities of the cyber domain. While this can seem straightforward at first, the real challenges will come when applying the law to real world cyber incidents. Future cyber attacks will become test cases for international law, and it is not a given that all members of an alliance or coalition will interpret the law in the exact same way.

While there might never be such a thing as 'cyber war', it is plausible that a cyber attack could *lead to* war. Article 2(4) of the Charter of the United Nations states that 'All Members shall refrain in their international relations from the threat or use of force against the territorial integrity or political independence of any state, or in any other manner inconsistent with the Purposes of the United Nations'. Michael N. Schmitt in a seminal article on the subject interprets the *use of force* in the context of the UN Charter as the use of *armed* force. This leads to the question of what constitutes arms in the cyber domain. To evaluate this, legal scholars look at the potential consequences of an attack, examining both indirect and direct effects. The political context within which an attack takes place is also of great importance to such an assessment (Schmitt 1999). If cyber means are considered arms and an attack is considered to have severe enough consequences, a state's use of cyber means against another state could be considered a breach of the prohibition of the use of force and could in turn lead to military operations as a response. A challenge in this respect is how to determine what a proportionate response to a given cyber attack would be, in order for a state to stay within international law in its actions. Are only responses within the cyber domain considered proportionate or can actors claim to be reacting proportionally while utilising kinetic means? Each future case where this is considered will be a test case for the current international law, and it is not a given that all parties to a conflict will reach the same conclusion when assessing the issue of proportionality.

There are two exceptions to the UN Charter's prohibition of the use of force: the right to self-defence and an explicit mandate from the United Nations Security Council (Waxman in Pedrozo and Wollschlaeger 2011). Most operations have an explicit mandate from the Security Council that determines what the mission ought to accomplish. It rarely describes in detail which means are to be used to achieve this. Mandates will to a large extent be shaped by the permanent members of the Security Council. For the cyber domain this becomes especially interesting since three of the permanent members – the United States, the Russian Federation and the People's Republic of China – are the three major players in cyberspace. These three major powers do not necessarily agree on how cyberspace should be utilised and where to draw the lines for accepted behaviour. While cyber issues are clearly high on the agenda for these three major powers, their approaches and attitudes towards the use and utility of the cyber domain differ. It is therefore not a given that these states will have a unified view on whether another state is overstepping their boundaries in cyberspace, much as is the case for other issues which are brought to the UNSC. Since Security Council mandates have grown wider, it seems unlikely that cyber operations will be mentioned explicitly in the text of a resolution calling for military efforts. It is, however, to be expected that cyber operations will to a larger extent be part of the escalation of a conflict and that the Security Council will issue resolutions condemning a state's use of cyber means either against its own population or other sovereign states.

One of the cardinal principles in international humanitarian law is the principle of distinction between civilian and military targets, between the civilian population and combatants. The principle also applies to cyber operations as part of a conflict. However, the UN Charter and Additional Protocols make it clear that it is *attacks* against civilians that are prohibited. As discussed above, whether the use of cyber means is considered an armed attack is not a clear-cut issue (Schmitt 2011). If attacks in the cyber domain are considered below the threshold for an attack according to *jus in bello*, a military cyber force could potentially attack civilians, making them especially vulnerable in a conflict situation. It is more difficult to separate the civilian and the military aspects of society in the cyber domain, as its infrastructure is both interconnected and interdependent. This makes it difficult even to define the operational environment for modern international operations when considering cyberspace as a war-fighting domain, let alone to distinguish between clearly defined civilian and military targets.

To sum up, when it comes to the cyber domain, existing international law is still valid. However, the interpretation of discrete cyber incidents will be a challenge for the future, as any incident will be a test case for the current legal regimes. In multinational forums it is far from obvious that all parties will reach the same conclusion.

Consent

When conducting international military operations, it is not only a matter of conducting them in accordance with domestic and international law. The forces also have to seek consent for their operations. The chapter on parameters describes how forces in modern international operations require consent not only from state authorities but also from non-state actors. This is perhaps even more prominent when considering the cyber domain, as most actors in cyberspace are non-state actors. When seeking consent from the general population, this section argues that the cyber domain becomes both a powerful tool and a battle space of its own as the Internet and social media 'involve' people in conflicts around the world in 'real time'.

As observed by Smith, states are never the sole parties in modern confrontations and conflicts. Modern military operations are fought by multinational coalitions against rebels groups, militias and traditional forces alike (Smith 2006). This is highly analogous to the cyber domain, as most actors in cyber space are non-state actors. This presents a new set of challenges for nations in dealing with cyber attacks, especially since attributing a cyber attack to a specific actor is extremely difficult. The aforementioned 'patriot hackers' present such a challenge for states on the receiving end of an attack, as these hackers are not controlled by the state, at least not officially, but can still get involved in conflicts through the cyber domain. A state cannot legally retaliate against a state assumed to be backing the attacks, while the actual individuals executing the attacks are likely to be regarded as criminals rather than combatants. It is, however, important to keep in mind that, while most actors are non-state, the most able, and therefore most dangerous, are traditional state actors (Windvik and Diesen 2013).

Smith (2006, p. 284) emphasises that we fight amongst the people also in a wider sense, namely through the media. A multinational force requires consent from the population at large as well as non-state actors. Groupings or individuals which do not accept or approve of what international forces plan or are mandated to do make use of the Internet both for cyber attacks and in the battle of the narrative. Smith highlights as one of the major trends of modern warfare that war takes place 'amongst the people, over the will of the people ... in seeking to establish conditions, our true political aim, for which we are using military force, is to influence the intentions of the people' (Smith 2006, p. 277). This did not come about with the introduction of cyberspace. However, the influence dimension, often known as information operations, has never been more relevant or had more opportunities than today. Information operations include a wide range of aspects, from cyber defence efforts to promoting the actions of a multinational force to the population in theatre (NATO 2009). There is, however, a distinction in principle between cyber operations based on illegitimate access to an adversary's system and information operations utilising the Internet and the opportunities it brings in a legitimate manner.

Television brought conflict closer to most people, as was evident as early as the Gulf wars in the nineties. However, since Smith first published his book in 2005, there have been considerable developments in media and technology. There has been an explosion in social media technology bringing conflict closer to people at a much more rapid pace, through services such as YouTube, Facebook, and Twitter. The home audience of nations participating in coalition operations abroad can be constantly updated online about developments in theatre, like Western populations following the developments in Afghanistan. This exacerbates challenges surrounding the legitimacy of operations in national constituencies – the conflict can be perceived differently by those following it from the safety of their own living rooms than by those in theatre.

The micro-blogging service Twitter has been described as the 'new battleground for NATO and Taliban in Afghanistan' after ISAF Public Affairs officers took to Twitter to counter false messages spread by Taliban spokesmen (CNN online 2011). The *Washington Post* describes the development, saying that the 'Twitter war' began in the midst of a sustained attack on the US Embassy and ISAF Headquarters in Kabul. 'How much longer will terrorists put innocent Afghans in harm's way', @isafmedia asked Taliban twitterer Abdulqahar Balkhi, who responded that ISAF had 'bn pttng thm n "harm's way" fr da pst 10 yrs' ['been putting them in "harm's way" for the past 10 years']. ISAF press officers say that this was the moment when the @isafmedia Twitter account went from being a passive tool to an engaging one to counter the propaganda issued by accounts such as @abalkhi, whose account is believed to have 'strong operational ties' (*Washington Post* online 2011b).

Technological developments have brought out-of-area operations into the lives of homeland populations on a real-time basis. The consequence is that a multinational force now needs increasingly to consider its own populations, in addition to the local population in theatre, when seeking consent.

Operational environment

Consent is, however, far from the only parameter which has been dramatically impacted by technological developments. Traditionally, the operational environment was defined as all the physical factors an intervention force is affected by. It has since been expanded to include the human terrain, as an intervening force needs to consider the local population. Kilcullen argues that the theatre of war, traditionally defined as areas 'of air, land and water that is, or may become, directly involved in the conduct of a conflict' is changing as a result of the emergence of the cyber domain (Kilcullen 2013, p. 171). Theatres of operations used to be described in terms of geographical areas where forces where physically located, while at present we see 'virtual theatres' that 'draw in populations and forces with no geographical connection to the conflict and which may be located anywhere on the planet' (Kilcullen 2013, p. 172). With the emergence of cyberspace, the logical layer (the logical connections between computers in a network) should also be included as part of the operational environment. The cyber domain is by design without boundaries and does not necessarily take into account geography and physical borders between states, a fact that will be explored in the following paragraphs.

When an intervening force is conducting offensive cyber operations against targets in the operational area, it needs to be particularly aware of cyber dependencies across national borders. Societies are highly intertwined and interdependent in cyberspace, and these networks are extremely complex. It is very difficult to chart the consequences of an attack with precision, and escalations into neighbouring areas are at times incredibly hard to predict (Clemente 2013). It is also important to consider the fact that the physical infrastructure which the logical layer rests on is owned by people, corporations or states. Taking out cyber-dependent services or assets in friendly or neutral states can do irreparable damage to the reputation of the international operation, in the same way as 'collateral damage'. Attacking, although unintentionally, assets that belong to friendly or neutral actors will reflect poorly on the entire multinational force (Johnsen 2013). This is especially relevant when considering the consequences of the aforementioned social media revolution, allowing news to travel unhindered across the world at a very rapid pace. This type of negative documentation in the form of videos and pictures shared via social media caused the Syrian regime to suspend access to the Internet for Syrians on 4 June 2011 (Kilcullen 2013, p. 222).

The consequence of virtual theatres is not only that it allows populations and forces far away from a war zone to participate despite physical distance; it also increases the risk of homeland attacks for nations contributing troops to international military operations. Cyber activists and patriot hackers do not have to attack an intervening force in the operational area; it might even be easier to attack civilian targets that are protected to a lesser degree 'back home' (Johnsen 2013). This is complicated further by the cyber domain enabling the execution of combat missions from bases on the other side of the world. Kilcullen refers to personnel taking part in the war in Iraq operating Predator and Reaper drones from Creech Air Force Base in Indian Springs, Nevada. The drone operators are classified as combatants, yet they live far from any war zone, in an area subject to United States domestic law. The same is true for those piloting remotely controlled aircraft from the suburbs of Syracuse, New York, targeting members of the Taliban located in Afghanistan and the tribal areas of Pakistan. While New York is physically located halfway around the world from Afghanistan, the leader of the Pakistani Taliban, Hakimullah Mehsud, attempted a suicide attack against New York City in 2010 in retaliation for Predator strikes against the Taliban which were controlled from the United States (Kilcullen 2013, pp. 172–176). The virtual theatre thus provides not only an advantage but also a possible increased risk for civilians in the homeland.

As technology has developed, the operational environment has developed with it. Theatres of war are no longer limited to a geographical war zone, as combatants can just as well be located on the other side of the globe. Cyber dependencies across national and military–civilian borders make it increasingly difficult to separate a conflict area from the rest of society.

Relative force composition and strength

As combatants and civilians alike are involved in conflicts through the use of cyberspace, it is easy to get the impression that nation-states have completely lost their competitive edge in the cyber domain. That is, however, not the case, as the cyber game is still one where the biggest players thrive.

Kilcullen describes a trend which he calls the democratisation of technology, a pattern which is categorised by the breaking down of 'classical distinctions between governments and individuals, between zones of war and zones of peace, between civilians and combatants and therefore between traditional concepts such as "war" and "crime" or "domestic" and "international".' (Kilcullen 2013, p. 176). This trend encompasses two separate but connected aspects: the democratisation of lethality, allowing individuals to access weapons that were only available to nation-states in the past; and the democratisation of digital connectivity, allowing individuals to access communication and control systems with far greater range than ever before (Kilcullen 2013, p. 176). To a large extent, this also holds true with regard to activities in cyberspace. Individuals and non-state actors currently have broad access to cyber 'weapons' and can achieve some level of 'operational effect' by utilising the cyber domain, and this is likely only to increase in the future. However, it is crucial to remember that the state actors which are greatest by conventional military might are the most powerful in the cyber domain as well. Stuxnet could, for instance, not have been developed and executed by any other actor than a highly technologically advanced nation-state with very sophisticated intelligence capabilities (Windvik and Diesen 2013; Healey 2013). The cyber domain is thus still a game in which the conventional major players remain in the lead. Generally, when speaking of the most powerful states in the cyber domain, one refers to the United States, Russia and China.

The United States has declared that should they fall victim to a major cyber attack, nothing is 'off the table', and that a potential retaliation may not only take place in the cyber domain (Stars and Stripes online 2009). The United States Cyber Command (USCYBERCOM) reached full operational capability in 2010, and the US Department of Defense named cyberspace a new domain of warfare in 2011. USCYBERCOM has three focus areas: 'Defending the DoD [Department of Defense] information networks, providing support to combatant commanders for execution of their missions around the world and strengthening our nation's ability to withstand and respond to cyber attack' (United States Strategic Command online 2013). The *Washington Post* claimed that US intelligence services conducted 231 offensive cyber operations in 2011, based on documents obtained from former National Security Agency (NSA) contractor Edward Snowden (*Washington Post* online 2013). The documents leaked by Snowden revealed extensive cyber surveillance by the NSA and caused international uproar. Snowden is wanted by the American government, but he left his native US for Russia immediately following the leaks.

As previously mentioned, Russia (or Russian patriots) have made use of the cyber domain in conflicts with its neighbours already but do not use the term 'cyber' when describing their own capabilities. Russia sees cyber means as part of a more holistic concept of 'information warfare', which includes intelligence, electronic warfare and information operations (Giles in Czosseck *et al.* 2012). While much of the Russian cyber focus has been on maintaining internal stability, especially with regard to limiting anti-government propaganda, Timothy L. Thomas claimed that 'nation-states should expect to encounter Russia's electronic presence on the virtual battlefield' (Thomas 2009, p. 487). Representatives of the Russian government announced in 2013 that the Russian Armed Forces would get their own cyber warfare branch, 'as the Internet could become a new "theatre of war" in the near future' and cyberspace is becoming a priority for Russia (RIA Novosti online 2013).

Chinese cyber capabilities and related intentions continue to grow increasingly visible. In February 2013, the cyber security company Mandiant issued a report which detailed the Chinese People's Liberation Army's 'direct involvement in hacking into American government and corporate websites' (Reuters online 2013). Media outlets termed the report a 'smoking cyber gun', and claimed it gave the 'most detailed look to date' inside the Chinese government's operations in cyberspace. According to Reuters, however, the Chinese Defence Ministry denied the accusations and called them 'unprofessional' (Reuters online 2013). China is also suspected of supporting campaigns of cyber attacks against neighbouring nations Japan and Taiwan as acts of retaliation. It is, however, not clear whether these attacks were results of patriot hackers or government efforts. China does not view the military and civilian spheres as discrete arenas, as is common in the West. Instead, China sees its general population as 'a major auxiliary information fighting force' (Thomas 2009, p. 471). In case of major conflict, the idea is that pre-recruited civilians with cyber skills can step in to reinforce the military cyber personnel in their efforts (Thomas 2009).

In summary, despite the democratisation of technology, large nation-states are still the most capable when considering cyber operations. If a force is conducting international military operations that one of the larger cyber powers is less than supportive of, that force should stay alert in terms of cyber defence efforts.

Conflict intensity

Cyber operations conducted on their own would be considered fairly low-intensity warfare. First, the 'cyber soldiers' do not even need to be present in theatre in order to partake in cyber operations, leaving them out of harm's way in terms of physical injuries (Kilcullen 2013). Second, as mentioned in the Stuxnet example, it is extremely rare to inflict physical damage through a cyber attack. In Rid's argumentation behind the statement 'cyber war will not take place', he claims that, for something to be classified as an act of war, it has to be 'potentially or actually lethal, at least for some participants on at least one side' of a conflict. And thus far, no cyber attack has reached this level of lethality. Therefore, he claims that 'all past and present political cyber attacks – in contrast to cyber crime – are sophisticated versions of three activities that are as old as human conflict itself: sabotage, espionage and subversion' (Rid 2013). However, when applying cyber means in concert with other means of force it can certainly contribute to a high level of conflict intensity.

Opponents of a large intervening force 'will seek to operate in ways which ameliorate their less sophisticated weaponry and organisation', something which has already become evident when considering cyber-space (Smith 2006). The United States describes its concern that adversaries have found ways to mitigate western military superiority by using so-called 'anti-access/area denial (A2/AD)' capabilities. By focusing on cyber capabilities that deny access to or freedom of movement within an area, adversaries are using an international force's dependency on the cyber domain against it, making cyber means a force multiplier (United States Department of Defense 2012).

Cyber means can be used in combination with other means of military force projection as part of a strategy to show an adversary the strength of force, showing that one can attack on multiple fronts at the same time. Cyber operations would in such a case be one of the building blocks in an effort to produce the effect of maximum violence (Diesen 2013). One example of this type of coordination is sketched out in Richard Clarke's book 'Cyber War': the Israeli Air Force was able to strike Syrian nuclear facilities in September 2007 without any aircraft appearing on Syrian radar screens, probably as a result of a combination of cyber means and electronic warfare (Clarke 2010, pp. 5–8; Morton in Healey 2013, p. 225). Coordinated operations across domains require a high level of synchronisation across both time and space. Sensors need to be turned off at the specific time that forces are moved physically and need to stay down for the duration of the move. This is very challenging (Windvik and Diesen 2013).

While conducting cyber operations coordinated with operations in the other warfighting domains is high in intensity, there are also additional challenges when conducting these operations coordinated with partner nations as part of a multinational force.

Cyber challenges for international military operations

This section will highlight a few interesting challenges pertaining to 'internationality' and the cyber domain. While the benefits of international military operations are many, in terms of access to more forces, increased legitimacy of numbers, the spreading of risk and affording multiple partners a seat at the table, there are also complicating features of operating together (Smith 2006). Future adversaries will use a coalition's dependency on the cyber domain against it, and in order to operate effectively with allies and conduct successful operations, interoperability stands out as a key feature. NATO defines interoperability as the ability of different military organisations to conduct joint operations (NATO online 2006).

NATO has only been granted a defensive mandate in cyberspace by the member states and has declared that cyber *defence* will continue to be a core capability of the Alliance. The NATO Cyber Defence Policy stresses that it seeks a coordinated approach to cyber defence which focuses on the prevention of cyber attacks and building resilience, as well as enhancing the Alliance's situational awareness capabilities. NATO's Strategic Concept also highlights NATO's goal of 'better integrating NATO cyber awareness, warning and response with member nations' (NATO 2013). In 2012, NATO suffered 2500 significant cyber attacks, according to NATO Review. These attacks probably consisted of everything from cyber attacks with the aim to disrupt, destroy or degrade NATO systems to intelligence and reconnaissance efforts. It was thereafter agreed at ministerial level that the NATO cyber defence capability should be at full capability by autumn 2013. NATO Secretary General Anders Fogh Rasmussen said,

> We are all closely connected. So an attack on one Ally, if not dealt with quickly and effectively, can affect us all. Cyber defence is only as effective as the weakest link in the chain. By working together, we strengthen the chain.
>
> (NATO Review 2013)

To do this, allies must seek to use interoperable solutions in the cyber domain as well.

The fact that a European Union force will be multinational by design is seen as a factor which could increase its cyber vulnerability. The EU aims to mitigate any cyber vulnerabilities resulting from the multinational nature of EU-led operations by increasing coordination among the national cyber defence capabilities of participating member states. To increase the effectiveness of this approach, the European Defence Agency is exploring opportunities to improve the civil and military cyber defence capabilities of EU member states, for instance through exercising together. The EU is also seeking to further industry cooperation among its member states, in order to provide interoperable solutions for the future (ENISA 2013).

Since NATO and the EU both have only defensive mandates in the cyber domain, there is increased potential for complications when conducting multinational operations, especially when considering synchronisation of effects in the cyber domain with those in the physical domains. Combined with the amount of secrecy surrounding offensive cyber operations, it is difficult to imagine operations being conducted without conflicting actions between partners. Clarke states that cyber operations are 'shrouded in such government secrecy that it makes the cold war look like a time of openness and transparency' (Clarke 2010, p. xi). There are also differing levels of controversy surrounding offensive cyber operations. In some countries, such as some of the Scandinavian ones, offensive use of cyber means is highly controversial, while in countries such as the United States the discussion is more frank, and the USCYBERCOM mission is to 'use information technology and the Internet as a weapon' (Clarke 2010, p. xi). As previously mentioned, the US has also declared that should they fall victim to a major cyber attack, nothing is 'off the table' (Stars and Stripes online 2009). This declaration can be viewed as an act of deterrence in cyberspace, convincing adversaries that the cost of attacking the US using cyber means is too great, as it could lead to a counter-attack with the full range of US military capabilities.

If an individual nation conducts cyber operations in the operational environment without informing its coalition partners of its intended targets and effects, there is a definite possibility of stepping on each other's toes, and unintentionally sabotaging partner operations by doing so. So-called 'blue-on-blue' incidents in the cyber domain, where for example one nation's cyber forces take out another nation's forces' mission critical systems resulting in injury or death, could as a result be damaging to alliance cohesion. How will this be avoided in the future, as forces grow even more 'cybered'? One option is treating it similarly to how the special operations forces are treated today, where they remain under national command throughout an operation, and have the opportunity to 'deconflict' with national forces in order to avoid unintentional sabotage. This is, however, more difficult in the cyber domain than in the physical realm, as it is more difficult to achieve a complete overview in advance with regard to unintended effects and consequences in cyberspace. It is also a complicating factor that the cyber domain transcends boundaries to a larger extent compared to special operations forces.

The United Nations has thus far mainly focused on the cyber domain at the political level rather than the operational. There has been a range of efforts on the Internet governance side but little emphasis on cyber defence and cyber operations in UN-led military operations. As the chapter on UN peacekeeping operations describes, troop contributions to UN operations have long been dominated by African and Asian contributors as Western nations have been tied up with ISAF in Afghanistan. If Western troops were to return as major troop contributors, this could pose complex challenges in the cyber domain. First, in terms of interoperability it would be more challenging to operate with African and Asian partners. Second, less technologically advanced nations might not put the same emphasis on cyber security and resilience. This could jeopardise potential cyber assets in multinational operations and make it more difficult to operate together.

While an alliance is of a more permanent nature, coalitions are usually ad hoc affairs. There is a trend in international military operations that coalitions of the willing spring out of groupings such as NATO. An example of this was Operation Unified Protector, a NATO operation where allies worked alongside partner nations both from their own geographical area, like Sweden, and partners from the region surrounding the theatre of operations, like Qatar. This adds another complicating factor to the conduct of operations in terms of being interoperable in the cyber domain. When partners like NATO or EU member states are reluctant to share information about their cyber capabilities and cyber operations with each other, one can only imagine how difficult this can become when operating with untraditional and new partner nations. Trust among coalition partners is important, especially when considering the sharing of information on sensitive topics such as cyber operations (Johnsen 2013). One way of countering this type of distrust is working on developing closer relationships with partners, by training and operating together, as the European Union is doing through establishing cyber training ranges (ENISA 2013).

To sum up, there is no shortage of cyber challenges for an international force conducting operations together. However, to counter some of these challenges it is important to strive for interoperable solutions both in terms of technology and procedures. It is also crucial to train and exercise together, both in order to operate better together and to forge trust among partners.

Conclusion

Cyber efforts are unlikely to be the sole means of force in future operations, but you will be likely to be hard pressed to find a future military operation where the cyber domain does not play a role. Cyberspace can be utilised in a wide array of different ways, such as in combination with other military means to show strength of force and ability to strike at multiple fronts at the same time. This can be done to influence the will of an adversary, with the aim of convincing an opponent to change its behaviour. In the early phase of a conflict, one can also make use of effects in the cyber domain with utility, as an attack in the cyber domain could be seen as less serious or aggressive than an attack in the physical domain. Cyberspace still represents a domain in which states can attack each other without causing physical harm to human beings.

However, the lack of escalation control is still a factor. The traditional military domains are thoroughly regulated and the consequences of most actions are well known, but in the cyber domain the perception of the attacked state is difficult to predict. There might also be differing perceptions of an attack within a coalition, which can damage coalition cohesion. Perception becomes crucial, as what is considered a minor attack by some can be interpreted by the attacked as a major breach of sovereignty, which could in turn lead to full-scale military confrontation. It is also difficult to predict unintended effects of the actions one takes in the cyber domain.

Should a party in a coalition have access to parts of the adversary's command and control systems, one option is to manipulate its information, for instance by changing map coordinates or timings in an operation order. This can lead to two outcomes: if the adversary does not notice that his coordinates have been changed, his operations can fail; if the adversary does notice, it can lead to widespread confusion and insecurity regarding which parts of the system are compromised and which can be trusted. This can influence his will to continue fighting, as well as his operational tempo. This type of access can also be used for intelligence, one of the main activities in the cyber domain. The dilemma here is whether to use access for intelligence purposes and stay hidden, or alter information and risk exposing your access. Finally, one of the most challenging forms of cyber operations is carefully planned and prepared operations aiming at the suppression of enemy sensors, information systems, etc. during a critical time slot, to enhance the effect of other weapon systems.

Perhaps less plausible uses of the cyber domain in military operations in the future are operations which lead to physical destruction of computer controlled infrastructure. Stuxnet is the only publicly known example of this ever happening, and it required nation-state level resources, several years of preparation, world-leading computer skills and finally that the infrastructure was designed in a way that enabled it to physically destroy itself. Taking out critical infrastructure for prolonged periods of time is also seen as extremely difficult.

The five parameters sketched out in Chapter 2 of this book are all impacted by the emergence of the cyber domain. The current international legal frameworks still apply; however each cyber incident on the international stage will serve as a test case for interpretation. War amongst the people has become even more relevant than when Smith published his book in 2005, as social media take any military operation into the pockets of individuals across the globe via their smart phones with Internet connections. Theatres of war are no longer only geographically based; they are virtual, and combatants can be placed on different continents yet still be striking their targets. This does not, however, completely change the relative strength on the world stage; large technologically advanced nation-states still have the advantage when it comes to complex and severe cyber attacks. Finally, while cyber operations on their own are fairly low intensity, the combination of cyber with other means of force can still amount to high-intensity warfare. There are also additional challenges when conducting operations along with partner nations as part of an international force. However, these can to some extent be countered by striving to achieve interoperable solutions both in terms of technology and procedures, and by establishing a level of trust between partners through training and exercising together.

While the technological developments we have seen over the last few decades have provided both additional challenges and advantages to conducting international operations, it is still more science fiction than reality to imagine future operations as bloodless cyber wars. However, it is safe to expect that any future war will have a cyber component.

Note

1 According to the NATO Review, there were 513 million Internet users in the world in 2001, which constitutes just over 8 per cent of the global population. In a little over a decade, this number had grown to 39 per cent of the world's population, with over 2.7 billion Internet users. However, the issue is not only that people use the Internet; people and businesses depend on it for conducting everyday tasks and providing services and thus for keeping up their way of life.

References

Arquilla, John and David Ronfeldt (1993). "Cyberwar is Coming!" *Comparative Strategy* 12(2) Spring 1993, 141–165.

Campen, Alan D. (1992). *First Information War: The Story of Communications, Computers and Intelligence Systems in the Persian Gulf War.* Fairfax, Va: AFCEA International Press.

Clarke, Richard A. and Robert Knarke (2010). *Cyber War: The Next Threat to National Security and What to do About it.* New York: Harper Collins.

Clemente, Dave (2013). *Cyber Security and Global Interdependence: What Is Critical?* London: Chatham House, The Royal Institute of International Affairs.

CNN online (2011). *Twitter is new battleground for NATO and Taliban in Afghanistan.* http://edition.cnn.com/2011/11/18/world/asia/afghanistan-twitter-war.

Diesen, Sverre (2013). "Cyber Operations – Game Changer or Supporting Activity?" Presentation at Norwegian Armed Forces Cyber Defence's Cyber Conference, 18 April 2013.

DR online (2012). www.dr.dk/Nyheder/Indland/2012/06/04/204156.htm.

ENISA (2013). *Mainstreaming European Military Cyber Defence Training and Exercises.* www.enisa.europa.eu/activities/Resilience-and-CIIP/cyber-crisis-cooperation/conference/2nd-enisa-conference/presentations/wolfgang-roehrig-eda-mainstreaming-european.pdf.

Giles, Keir (2012). "Russia's Public Stance on Cyberspace Issues". In C. Czosseck, R. Ottis and K. Ziolkowski, Eds. *4th International Conference on Cyber Conflict.* Tallinn: NATO CCD COE Publications.

Healey, Jason, Ed. (2013). *A Fierce Domain: Conflict in Cyberspace 1986 to 2012.* Washington DC: Atlantic Council and Cyber Conflict Studies Association.

Johnsen, Siw Tynes (2013). "Military Operations and the Cyber Domain". Lecture at the University of Oslo, 22 October 2013.

Kilcullen, David (2013). *Out of the Mountains – The Coming Age of the Urban Guerilla.* London: Hurst and Company.

Kramer, Franklin D., Stuart H. Starr and Larry K. Wentz, Eds. (2009). *Cyberpower and National Security.* Washington, DC: Center for Technology and National Security Policy, National Defense University Press.

Morton, Chris (2013). "Olympic Games". In Jason Healey (2013). *A Fierce Domain: Cyber Conflict 1986 to 2012.* Washington DC: Atlantic Council and Cyber Conflict Studies Association.

NATO (2006). *Backgrounder: Interoperability for Joint Operations.* www.nato.int/nato_static/assets/pdf/pdf_publications/20120116_interoperability-en.pdf.

NATO (2009). *Allied Joint Doctrine for Information Operations AJP-3.10.* http://info.publicintelligence.net/NATO-IO.pdf.

NATO (2013). *NATO and Cyber Defence.* www.nato.int/cps/en/natolive/topics_78170.htm.

NATO Review (2013). *The History of CyberAttacks – A Timeline.* www.nato.int/docu/review/2013/Cyber/timeline/EN/index.htm.

National Public Radio online (2013). *Pentagon Goes on the Offensive against Cyber Attacks.* www.npr.org/2013/02/11/171677247/pentagon-goes-on-the-offensive-against-cyber-attacks.

New York Times online (2011). *US Debated Cyberwarfare in Attack Plan on Libya.* www.nytimes.com/2011/10/18/world/africa/cyber-warfare-against-libya-was-debated-by-us.html?_r=0.

NRK online (2012). *Afghanistan-styrkene hacket.* www.nrk.no/verden/hacket-forsvarets-afghanistan-nett-1.8178990.

Reuters online (2012). www.reuters.com/article/2012/04/27/net-us-afghanistan-taliban-hacking-idUSBRE83Q09I20120427 Reuters online (2013). www.reuters.com/article/2013/02/23/net-us-hackers-virus-china-mandiant-idUSBRE91M02P20130223.

RIA Novosti online (2013). *Russian Military Creating Cyber Warfare Branch.* http://en.ria.ru/military_news/20130820/182856856.html.

Rid, Thomas (2013). *Cyber War Will Not Take Place.* London: Hurst/Oxford University Press.

Schmidt, Andreas (2013). "The Estonian Cyberattacks". In Jason Healey, Ed. *A Fierce Domain: Conflict in Cyberspace 1986 to 2012.* Washington, DC: Atlantic Council and Cyber Conflict Studies Association.

Schmitt, Michael N. (1999). "Computer Network Attack and the Use of Force in International Law: Thoughts on a Normative Framework". *Columbia Journal of Transnational Law* Vol. 37, 1998–1999. SSRN: http://ssrn.com/abstract=1603800.

Schmitt, Michael N. (2011). "Cyber Operations and the Jus in Bello: Key Issues". *Naval War College International Law Studies.* SSRN: http://ssrn.com/abstract=1801176.

Smith, Rupert (2006). *The Utility of Force: The Art of War in the Modern World.* London: Penguin Books.

Stars and Stripes (2009). *Official: No Options 'off the table' for US Response to Cyber Attacks.* www.stripes.com/news/official-no-options-off-the-table-for-u-s-response-to-cyber-attacks-1.91319#.UzqW2PmSxzg.

Thomas, Timothy L. (2009). "Nation-state Cyber Strategies: Examples from Russia and China". In Franklin D. Kramer, Stuart H. Starr and Larry K. Wentz, Eds. *Cyberpower and National Security.* Washington, DC: Center for Technology and National Security Policy, National Defense University Press.

United Nations (1945). *Charter of the United Nations.* www.un.org/en/documents/charter/index.shtml.

United States Computer Emergency Response Team (2013). *Understanding Denial-of-Service Attacks.* www.us-cert.gov/ncas/tips/ST04–015.

United States Department of Defense (2012). *United States Joint Operational Access Concept.* www.defense.gov/pubs/pdfs/joac_jan%202012_signed.pdf.

United States Strategic Command (2013). *Factsheet: Cyber Command.* www.stratcom.mil/factsheets/2/Cyber_Command.

Washington Post online (2011a). *Operation Cupcake: MI6 Replaces al-Qaeda Bomb-making Instructions with Cupcake Recipes.* www.washingtonpost.com/blogs/blogpost/post/operation-cupcake-mi6-replaces-al-qaeda-bomb-making-instructions-with-cupcake-recipes/2011/06/03/AGFUP2HH_blog.html.

Washington Post online (2011b). *US military, Taliban use Twitter to Wage War.* http://articles.washingtonpost.com/2011–12–18/world/35284991_1_isafmedia-abalkhi-social-media.

Washington Post online (2013). *US Spy Agencies Mounted 231 Offensive Cyber-operations in 2011, Documents Show.* www.washingtonpost.com/world/national-security/us-spy-agencies-mounted-231-offensive-cyber-operations-in-2011-documents-show/2013/08/30/d090a6ae-119e-11e3-b4cb-fd7ce041d814_print.html.

Waxman, Matthew C. (2011). "Cyber Attacks as 'Force' under UN Charter Article 2(4)". In Raul A. "Pete" Pedrozo and Daria P. Wollschlaeger *International Law and the Changing Character of War.* International Law Studies, Volume 87.

Windvik, Ronny and Sverre Diesen (2013). *Cyber Power and the Role of the Armed Forces.* FFI-Forum, Oslo Militære Samfund, 26 November 2013.

11 Protection of civilians as a new objective in military operations

Alexander William Beadle

Protection of civilians has emerged as a central objective in many of today's military operations. Traditionally, protection of civilians has been understood in terms of avoiding 'collateral damage', respecting the law of armed conflict and assisting with the delivery of humanitarian aid. Today, military forces are increasingly expected to protect civilians from perpetrators who deliberately target them and are responsible for the vast majority of civilian casualties. This has posed a new challenge in most operations, which remains largely unresolved and requires novel thinking about the utility of force. Protection of civilians will continue to be important in most future military operations, but the ways in which civilians are targeted and the ways in which they can be protected will be influenced by trends, such as urbanisation, proliferation of modern weapons and the availability of new technologies.

This chapter is divided into three parts. The first part explains the emergence of protection of civilians as an objective for military forces in operations today. It argues that the unprecedented importance attached to this particular objective can be attributed to changes in the nature of warfare itself rather than merely normative developments alone. In today's 'wars amongst the people', civilians have assumed the centre stage as both targets of deliberate attacks and objectives to be won, which has made their protection a key objective for both moral and military–strategic reasons. Despite this, there is currently an 'implementation gap' in all types of operations, whereby the strategic importance of protecting civilians has not yet been matched with an ability to do so on the ground. This has largely been attributed to a lack of sufficient thinking and guidance on how to operationalise the objective.

The second part of this chapter argues that this 'gap' conforms to General Rupert Smith's general observation about how military forces have failed to find utility in today's operations. By applying Smith's work on the utility of force and war amongst the people to the particular objective of protecting civilians, this chapter discusses how finding utility of force to protect civilians requires a different approach than is currently found in existing military doctrines for other types of operations. The key is to understand why and how perpetrators attack civilians, which is the very reason why protection becomes an objective in the first place. It is argued that we cannot find utility of force to protect civilians without understanding how perpetrators have found utility of force by attacking them in the first place. This permeates how each of the parameters (*mandate, conflict intensity, operational environment, consent,* and *relative force composition and strength*) must be understood from a protection perspective.

The third and final section looks beyond the current implementation gap and focuses on how the operationalisation of protecting civilians may change in the future. The main argument is that protection of civilians is likely to remain an important objective as long as war is fought amongst and against people. The impact of global trends will be most noteworthy in terms of the means and methods through which perpetrators and intervening forces will attack and protect civilians in the future.

The emergence of protection of civilians

Protection of civilians is today broadly defined as 'efforts to protect civilians from physical violence, secure their rights to access essential services and create a secure environment for civilians over the long-term' (Giffen 2010, p. 8). However, different actors have their own understanding of what is meant by protection of civilians – and its meaning has evolved over the years. Traditionally, the majority of protection activities have been conducted by humanitarian actors working to alleviate the immediate suffering of civilians in times of war, such as through the delivery of aid, long-term development activities and advocacy of human rights.

During military operations, the concern with protecting civilians has historically stemmed from legal conventions, treaties and regulations outlined in the law of armed conflict. These include the Geneva (1864, 1906, 1929 and 1949) and Hague conventions (1899, 1907) and subsequent treaties, case law and customary international law. These laws focus on obligations that armed actors have in relation to so-called 'protected persons' (such as the local civilian population, prisoners of war and injured soldiers) and impose restrictions on the conduct of war according to the four core principles of distinction, military necessity, unnecessary suffering and proportionality. What is relatively new is a greater readiness to use military force against armed actors that target civilians deliberately, which has only recently become a regular task for military forces.

The idea of military intervention for humanitarian purposes was largely absent until the end of the cold war. Three previous instances which could have been justified on a humanitarian basis – India in East Pakistan (1971), Vietnam in Cambodia (1978), and Tanzania in Uganda (1979) – were all condemned by the UN. Since then, normative developments have made intervention to 'save strangers' on humanitarian grounds increasingly legitimate (Wheeler 2000). This was in large part due to the failures of the international community to protect civilians in Somalia, Rwanda and Bosnia in the early 1990s. For the UN, protection of civilians has become an end in itself for moral reasons, as it cuts to the core of the organisation's very purpose to save people from the scourge of war. Hence, protection of civilians is today a top priority in most UN peace-keeping operations, where there is now a much greater readiness to use military force in order to match these ambitions with necessary action, as discussed in more detail in Kjeksrud's previous chapter.

The failures of the 1990s also highlighted the broader problem of pre-venting mass atrocities from happening in the first place. In 1999, NATO intervened in Kosovo to stop the ethnic cleansing of Albanians by Milose-vic's Serbian forces. This operation was considered 'legitimate' because it sought to save civilians under imminent threat, but also 'illegal' because no UNSC resolution authorising the use of force was ever passed. As a result, the International Commission on Intervention and State Sover-eignty was established in 2001. Its mission was to find out how gross and systematic violations of human rights could be stopped, while preserving state sovereignty. The result was the R2P principle, endorsed by UN member states at the World Summit in 2005. This principle remodelled the idea of state sovereignty from strict non-interference to include a state's responsibility to protect its own population from mass atrocities. When states fail to do so, or are attacking civilians themselves, this respons-ibility is transferred to the international community, with the use of military force as a last resort. The principle was used by the international community during the post-election clashes in Kenya (2008), where diplo-matic action under R2P is said to have prevented a plunge into deadly ethnic conflict (Weiss 2010). When NATO again intervened to protect civilians in Libya (2011), it was the first time R2P was used to authorise military intervention. As it happened, the gradual expansion of targets drew criticism from Russia in particular, who felt that the initial mandate to protect civilians had been hijacked and used to enforce regime change in Libya.

However, it is important to note the fundamental differences between R2P and protection of civilians, especially since this controversy surround-ing the use of R2P in Libya is said to have jeopardised the prospect of military intervention to protect civilians in the future. R2P and protection of civilians share a normative foundation – 'the protection of individuals' – but they concern fundamentally different aspects of achieving it (Global Centre for the Responsibility to Protect 2012). R2P is more about 'when' and 'under what conditions' it is right to intervene to save civilians from the four gravest violations of human rights (genocide, war crimes, ethnic cleansing and crimes against humanity); while protection of civilians con-cerns 'how' civilians on the ground can actually be made safer and better protected, regardless of the reasons for launching an operation. Protec-tion of civilians has been a legal obligation for decades and is what the

International Committee of the Red Cross (ICRC) and humanitarian organisations do on a daily basis. It will also remain a priority in the UN's peacekeeping operations as a thematic subject abbreviated to 'PoC', regardless of what happens to R2P.

In fact, the real concern for protecting civilians comes not merely from normative trends, but is a result of changes in the nature of warfare itself. Of the vast literature on modern warfare, General Smith's work on 'war amongst the people' is particularly relevant to protection of civilians because he combines the two central issues at stake – the role of civilians, and the use of military force. According to Smith, war amongst the people has replaced industrial warfare as the dominant paradigm in which today's conflicts take place. This is both a graphical and conceptual description of a new reality in which both civilians and military commanders find themselves:

> It is the reality in which the people in the streets and houses and fields – all the people, anywhere – are the battlefield. Military engagements can take place anywhere: in the *presence* of civilians, *against* civilians, in *defence* of civilians. Civilians are the targets, objectives to be won, as much as an opposing force.
>
> (Smith 2006, pp. 3–4, my emphasis)

For civilians, the reality has become one in which they are more likely to be killed than combatants are. Although the total number of civilians killed has declined because there are fewer major conventional wars, the ratio of civilian-to-soldier deaths has been inversed. According to a 2001 study by the ICRC, nine soldiers were killed for every civilian death during World War I, while it is now estimated that 'ten civilians die for every soldier or fighter killed in battle' (Greenberg and Boorstin 2001, p. 19). For individual conflicts, the ratio has been put at 3:1 in Afghanistan (Kemp 2011), 8:1 in Iraq (Iraq Body Count 2010) and 100:1 in the DRC (Lidow 2010, p. 2). These figures, however, only reveal the number of dead. Civilians are even more frequently subjected to rape, abduction, plunder and exploitation or harmed in other ways by an abundance of state and non-state actors. This situation has been the principal driving force behind the UN's efforts to make protection of civilians a priority in its own workings, primarily for moral reasons.

For military commanders, the fact that wars now take place amongst, against and in defence of civilians has major implications for how force may or may not be used. While decisive military victory could determine the outcome of conflicts in the past, such as by capturing territory or overthrowing a government, Smith argues that military forces today can only be used to establish the conditions for resolving the conflict by other means. In Smith's words, 'if a decisive strategic victory was the hallmark of interstate industrial war, establishing a condition may be deemed the hallmark of the new paradigm of war amongst the people' (2006, p. 270). Only by capturing the 'will of the people' may we ultimately achieve the outcomes we seek:

> In seeking to establish conditions, our true political aim, for which we are using military force, is to influence the intentions of the people. This is an inversion of industrial war, where the objective was to win the trial of strength and thereby break the enemy's will. In war amongst the people the strategic objective is to capture the will of the people and their leaders, and thereby win the trial of strength.
>
> (Smith 2006, p. 277)

It is this military–strategic rationale that has pushed considerations about protecting civilians into operations whose primary objective may be something else entirely, such as the counter-insurgency operations described in Chapter 4. Here, protecting the local population is viewed as a principal means through which to defeat an insurgency – not as an end in itself. This rationale is based on historical principles of counter-insurgency operations and recent experiences from Iraq and Afghanistan that illustrate the costs of *not* protecting civilians. First, there is the 'insurgent math', which holds that every civilian casualty forces cause themselves will create an additional 20 insurgents (*Army Times* 2010). Second, civilian casualties buttress the insurgents' propaganda. Finally, the general lack of trust it generates amongst the population, whose primary concern is their personal safety, is a real problem as far as winning hearts and minds is concerned. Thus, ISAF's guidance in Afghanistan explicitly stated in 2009 that 'protecting the people is the mission' (ISAF Counter-insurgency Guidance 2009, p. 1). Several measures have since been implemented to reduce the number of civilian casualties caused by own actions, such as tactical directives restricting the use of close air-support, escalation of force and the establishment of civilian casualty tracking cells (Beadle 2010). These measures have contributed to a reduction in the proportion of civilian casualties caused by pro-government forces in Afghanistan from 41 per cent in 2007 to 11 per cent in 2013 (UNAMA 2009; UNAMA 2014). The vast majority of civilians are now killed by the Taliban and other insurgent groups.

Yet civilians have not become better protected in many, if not most, operations. While international and government forces are responsible for fewer civilian casualties in Afghanistan, the total number of civilians killed doubled from 1,523 in 2007 to 3,021 in 2011 (UNAMA 2009, 2012). Since then, the number of civilian deaths has remained around 3,000, although the number of injured has continued to increase (UNAMA 2014). For all its good intentions, the UN too continues to struggle with its principal mandate to protect civilians – being criticised both for using 'too little' force in most situations and 'too much' in a few others. For example, the

UN was in 2009 criticised for its support to the Congolese armed forces that allegedly killed hundreds of civilians during an offensive against armed groups in eastern DRC, whilst it was in late 2012 criticised for not doing the same as M23 rebels advanced on Goma. Thus, despite their fundamentally different reasons for doing so, the inability to protect civilians from perpetrators of violence remains a principal concern and challenge for both the UN and NATO (Beadle 2010).

Finding 'utility of force' to protect civilians in today's operations

This continuing failure to protect civilians in many, if not most, contemporary operations is not surprising. Precisely because protection of civilians from perpetrators of violence represents a relatively new task for military forces, there are no established military theories, doctrines or concepts to draw upon. Repeated failures in the past have shown that military doctrines and concepts designed for other types of operations are unable to provide guidance on how to actually go about protecting civilians on the ground (see Giffen 2010; Sewall *et al.* 2010; PKSOI 2013). The result has been an 'implementation gap' shared by most actors, whereby civilians have not been better protected despite the unprecedented strategic importance attached to this objective by all actors (Beadle 2010).

This struggle to find ways in which force can be used to protect civilians conforms to Smith's general observation of how we fail to use military force in ways that help establish the conditions necessary to achieve our end states. It is this basic problem of finding 'utility of force' in today's wars amongst the people – 'how to use armed force to achieve a desired and stable political outcome' – that Smith takes issue with (Gow 2006, p. 1161).

In order to find better utility of force in today's wars amongst the people, Smith has called for a renewed analysis of how political ends and military means can be reconciled to achieve the conditions we seek to establish. 'Protection of civilians' is likely to be one such condition in *any* of today's wars amongst the people, because it is regarded as a prerequisite for achieving other objectives, as well as an end in itself for moral reasons. Military forces are nearly always expected to protect civilians under imminent threat of physical violence, even in operations that are not explicitly 'mandated' to do so. When military force is deployed or employed in such a way that it successfully lowers this threat of physical violence, what can be referred to as 'utility of military force to protect civilians' has been found (Beadle 2011).

Finding such utility, however, requires a basic understanding of the nature of protection of civilians as an objective. The criteria for assessing the utility of military force to protect civilians are different from assessing the utility of military force in defeating armed actors, establishing democracy or conquering new territory. On the one hand, the use of force must be able to reduce the threat of physical violence posed by a perpetrator to a lower level than it currently is at. At the same time, the use of force must not endanger more people than it protects in the process. This principle of 'do no harm' is derived from medical ethics and holds that it may be better *not to do something*, or even to do *nothing at all*, than to risk causing more harm than good. This goes beyond merely avoiding excessive 'collateral damage' as a result of own operations. The biggest threat to civilians comes from perpetrators who may intensify violence against civilians in response to certain military actions – or the lack of them.

Neither of these criteria, however, can be assessed without prior knowledge of the level of threat to civilians in the first place. Only then is it possible to determine the condition required to reduce it and what will happen if no action is taken at all. This challenges the ways in which we normally conceive of military operations.

For instance, the traditional parameter of 'conflict intensity' cannot alone be defined by the threat posed to our own forces but must also be defined by the intensity of threat posed to civilians by the perpetrator. Yet there are enormous variations in terms of how much and in what ways violence is used against civilians. Civilians may be under threat of being massacred, expelled, randomly targeted, selectively killed, brutally punished or looted. Each of these threats will require different military responses to reduce it.

To capture the range of different situations where civilians are faced with fundamentally distinct types of physical threats, the FFI has developed seven generic planning scenarios as a basis for determining what military responses may work in which situations (Beadle 2014):

- genocide, where perpetrators seek to *exterminate* a certain group, usually through multiple mass-killings at different locations simultaneously (e.g. Rwanda, 1994);
- ethnic cleansing, where perpetrators seek to *expel* a certain group, usually through demonstrative acts of violence and destruction of homes (e.g. Bosnia, 1992–1995);
- regime crackdown, where regimes use violence to *control* a restless population, usually through a gradual escalation from mass-detention, rape and torture to massive destruction of populated areas and occasional massacres (e.g. Libya, 2011);
- post-conflict revenge, where individuals or mobs *take revenge* for past crimes, usually through criminal acts like murder, arson, kidnapping and looting (e.g. Kosovo post-1999);
- communal conflict, where whole communities seek both to *avenge a previous round of violence* and to *deter further retaliation* as a means of protecting themselves, through raids, killings and destruction of means of survival (e.g. tribal conflict in Jonglei state, South Sudan, 2009 onwards);

- predatory violence, where perpetrators *exploit* civilians to survive or for profit, usually through plunder, taxation, forced recruitment, opportunistic rape and severe brutality against 'easy targets' (e.g. Sierra Leone, 1991–2002);
- insurgency, where insurgents target civilians merely as a way to *control* populations upon which they depend and *undermine the control of others*, usually through a combination of selective violence against collaborators and indiscriminate explosive attacks in areas under government control (e.g. Afghanistan, 2006 onwards).

All of these scenarios have been identified on the basis of the historical rationales that perpetrators have had for targeting civilians. While the particular tactics and methods used have varied greatly across time and space, these underlying rationales (e.g. to exterminate, to expel or to control a population) have remained fundamentally the same throughout history.

Understanding a particular perpetrator's motivation for attacking civilians will also help break down the complexity of other parameters, such as the 'operational environment'. Since protection of civilians concerns populations rather than territory, protection will always be most needed in population centres where potential victims are concentrated. However, *where* such protection 'hot-spots' may arise will vary according to the particular motivation of a perpetrator. For example, perpetrators seeking to exterminate or expel a certain group will have to target specific areas where that population resides. Thus, the Bosnian Serbs *had to* capture the Muslim enclaves of Žepa, Goražde and Srebrenica in order to achieve their goal of an ethnically pure contiguous territory. By contrast, insurgents, who only attack civilians as a way of undermining trust in the government, are far less dependent on attacking civilians in a particular location or at a particular time to achieve their intended effect.

In situations where civilians are deliberately targeted, the degree to which these perpetrators will violently oppose an intervening force will also vary and may be more assumed than real in some situations. There is an assumption that intervening forces will be attacked if they go in to protect civilians, which is based on experiences from Somalia, Iraq and Afghanistan. But all of these cases were situations where the perpetrators were non-state actors (warlords, terrorists or insurgents) unable to formally oppose any deployment and whose primary targets were government forces and their supporters, not civilians per se. By contrast, in those scenarios where civilians are primary targets (genocide, ethnic cleansing, regime crackdown, post-conflict revenge, communal conflict and predatory violence), an intervening force is less likely to meet heavy opposition, precisely because civilians are the primary targets against which the perpetrator must concentrate effort in order to achieve the goal. Such perpetrators, who are often quite militarily capable, have historically limited attacks on intervening forces to testing their resolve. The hope is that it will prompt their withdrawal, such as with the attack on Belgian peace-keepers in Rwanda and the taking of UN forces as hostages in Bosnia and Sierra Leone. This is not to say that an intervening force will not be opposed – but actual interference is less likely when an armed actor is primarily at war with his own population.

This means that we have to think differently about 'force composition and strength' as well. For starters, the traditional force ratio of 3:1 between attackers and defenders will not be valid for the relationship between armed perpetrators and unarmed civilians. Very small numbers of perpetrators can threaten large portions of the population in all scenarios. For instance, fewer than 600 LRA fighters are presumed to have been responsible for 1,200 killings of civilians, 1,400 abductions, displacing 200,000 people and destroying or looting thousands of buildings in the course of six months during 2009 (UNJHRO 2009, p. 24).

The relationship between perpetrators and protectors may also have to be assessed differently. In the aftermath of the genocide in Rwanda, a study found that a force of 5,000 well-equipped troops could have changed the outcome significantly. The most likely form of opposition was judged to be 'periodic blocking and interference by small organized units' of Tutsi rebel forces and the Hutu genocidal government they were fighting (Feil 1998, p. 10). In fact, the 3,000 mostly French forces that were deployed in June 1994 met very limited resistance. In Bosnia, the Bosnian Serb leadership judged in 1993 that 'their cause would effectively have been lost' if 5,000–10,000 NATO troops were deployed in the northern corridor linking the two Bosnian Serb areas (Gow 1997, p. 252; Silber and Little 1997, p. 279), while military planners in Europe and the US were contemplating force figures of between 75,000 and 400,000 to implement the Vance–Owen plan (Christensen 2011).

Very few examples of ground forces successfully protecting civilians from perpetrators exist. In 2003, the EU launched an operation (Artemis) to Bunia, DRC, to stabilise a situation where large-scale attacks on civilians were occurring and to reinforce a failing UN presence. The force of 1,850 mostly French troops successfully lowered the threat to civilians by implementing weapons-free zones in and around Bunia and by hitting back hard against militias terrorising the local population. More recently, the French deployment to Bangui in the Central Africa Republic is likely to have stemmed further escalation of the communal conflict there, even though it has not been able to reduce the desire for revenge and perception of threat on both sides. What is notable from these examples, and should be studied further, is how these figures stand in sharp contrast to the huge force numbers needed to protect civilians in situations where they are *not* primary targets, like classic insurgencies. Other studies on ratios in peacekeeping operations have found that fewer internally displaced persons per peacekeeper and square kilometres for peacekeepers to patrol have been associated with successful mitigation of violence (Ford Institute for Human Security 2009).

As for 'consent' from the victim populations, they will initially be supportive of any efforts to protect them if under threat – but their continuous consent is dependent on the ability of protectors to deny the perpetrators the ability to attack them. It will be impossible for military forces to protect everyone, but obvious 'protection failures' are the quickest way to lose consent. Even in operations where the primary objective is not protection of civilians, failure to do so when a threat arises is likely to cause a loss of legitimacy for the entire operation as a whole – both in-theatre and at home. If one does fail, civilians will at best ignore you, as they did in Bosnia, or may at worst fight you, as happened in Somalia and areas of Afghanistan.

That said, the actual tasks required to protect civilians would not necessarily be very different from traditional military activities. These include patrolling, checkpoints, observation posts, mobile operating bases, cordon and search operations, interposition operations, force protection and neutralisation of adversaries (see PKSOI 2013). Thus, the composition of a force designed to protect civilians may not be so different from a force deployed to the same operational environment with a different mandate (protection is rarely the *only* objective). What is different is when, where and how these military forces will be used.

This prompts a return to the original question facing all operations where protection of civilians is an objective today: how can force be used to protect civilians with greater utility? Smith has identified 'only four things' that military forces can do in wars amongst the people. At minimum, military forces can be used to 'ameliorate' a crisis, which does not involve the use of force but for instance the delivery of aid, construction of camps and other activities in aid of civilian life. The second function military forces may have is to 'contain' a situation, which involves only limited use of force to prevent a crisis from spreading or escalating, e.g. by enforcing arms embargoes or imposing no-fly zones. Third, military force may be used to 'deter' or 'coerce' an armed actor to form or change that party's intentions, by threatening or actually using force. At most, force may be used to 'destroy' an opposing force's ability to achieve their goals, which is what has traditionally been regarded as the primary purpose of military forces (Smith 2006, pp. 320–321).

Which of these particular functions of force will be most useful in terms of protecting civilians ultimately depends on how perpetrators attack civilians in the first place. The criterion for success in protecting civilians is that the use of force must lower an existing threat, without causing more harm than otherwise would have occurred. Thus we may only determine 'how' utility of force to protect can be found on the basis of how perpetrators have found utility of force by attacking civilians first. This is contrary to how operations have been designed in the past.

Traditional military theory tends to emphasise two *warring* parties – whether they are conventional or irregular military forces. According to the military theory on coercion, the will and capability of an adversary is assessed against *one's own* will and capability in order to determine how they may most effectively be compelled to do our will (Johnson *et al.* 2002). However, such an approach fails to consider the same actor's will or capabilities to *attack civilians*, who may be more concerned with expelling a certain group of people rather than fighting an intervening force. Furthermore, the operational requirements that an actor requires to attack civilians may be entirely different from those needed to resist another armed force. Clausewitz's famous dictum that the enemy's power is the sum of his available means and the strength of his will still rings true, but it will have to be assessed differently if protection is the objective for the use of force.

With protection of civilians, the starting point is an armed actor against unarmed civilians with little or no ability to defend themselves. While their will to survive may be great, civilians will by definition lose the battle of strengths against an armed actor. This is what prompts protection from perpetrators to become an objective for third-party military forces in the first place. The role of an intervening force is to even out this imbalance and, if possible, influence the perpetrator's willingness to continue targeting civilians.

The failure to recognise the true war aims of Serbian war efforts in former Yugoslavia (to carve out a contiguous and ethnically pure Serbian area of Bosnia and Croatia) meant that military force was used merely to protect the delivery of humanitarian aid ('ameliorate') and enforce a no-fly zone and arms embargo ('contain'). In reality, the threat to civilians could only be lowered through 'coercion' of the Bosnian Serbs. By the time this was done through a NATO air offensive in 1995, most of the ethnic cleansing had already finished. This shows how the functions of one's own use of force must, as far as it is possible, seek to match the perpetrator's use of violence against civilians in the first place (Beadle 2011, 2014).

Another potential source of failure is that the use of force does not degrade the perpetrator's ability to attack civilians. A case in point was NATO's operation in Kosovo, Operation Allied Force (1999), where thwarting ethnic cleansing of the Albanian population was one of three strategic objectives. The operation was designed to coerce Milosevic into stopping his attacks and forcing him back to the negotiating table. This involved striking targets that would gradually weaken him militarily and domestically, such as air defence, ammunition depots, command and control, economic and political infrastructure (factories, bridges, oil refineries) and assets of Milosevic's key supporters like factories and Serb targets in Vojvodina (an entirely different region). However, these targets were selected on the basis of Serbian will and military capabilities vis-à-vis NATO rather than the Serbian will or capabilities to ethnically cleanse the Albanians. In fact, even the Yugoslav army only played a supporting role by establishing firepower dominance and operating checkpoints, while small units of interior police forces armed and equipped as light infantry conducted the actual cleansing. Regular forces did play a role, but lessons from Bosnia indicated that ethnic cleansing could be achieved with light infantry alone, with only slightly greater risk to the units involved (Vick *et al.* 2001).

Thus, many of the targets struck were barely linked to Milosevic's ability to attack civilians, including the exaggerated destruction of Yugoslav military units. In fact, the ethnic cleansing of Albanians was perhaps the most successful aspect of Milosevic's strategy during the war – the others being breaking Alliance cohesion, retaining Russian support, and defeating the Kosovo Liberation Army. Although Milosevic eventually conceded, close to a million Albanians were expelled from Kosovo by Serbian forces and 90 per cent were displaced from their homes (Human Rights Watch (HRW) 2001, p. 4). Thus, the traditional theory of coercion is one example of 'enemy-versus-us'-focused military theories that cannot answer the basic questions of what military force can do to protect civilians. Only by assessing the perpetrator's use of violence against civilians will one be able to identify 'against what' military force can be used with purpose, what kind of forces will be required and how they may best be used.

At present, there are ongoing processes to address this 'gap' in thinking about how protection of civilians can be operationalised. These include a planning handbook intended for the US military on how to stop genocide and mass atrocity situations (Sewall *et al.* 2010), military principles for the protection of civilians (Kelly 2010), proposed guidelines for military planning to protect civilians in UN operations (Kelly and Giffen, 2011), a 'doctrine-like' military reference guide that takes a comprehensive approach and lists over forty different protection-related tasks (PKSOI 2013), and a step-by-step guide for military planners that provides advice on what may work in which particular situations based on the scenarios described above (Beadle and Kjeksrud 2014). What all of these efforts have in common is a profound emphasis on understanding the perpetrators of violence. A key finding has been that the role of military force is often greater than assumed, precisely because violence against civilians constitutes such an intrinsic part of the strategy of most perpetrators that non-military instruments of power are unlikely to be able to reduce the resulting threat.

Efforts to address this gap are likely to continue for years ahead, as new lessons are drawn from new operations. This also raises the questions of what may change in terms of protection of civilians in future operations and what is likely to stay the same.

Protection of civilians in future military operations

Just as with any other mission type or task discussed in this book, the nature of protection of civilians as an objective for military forces will also

change in light of the broader trends. The different rationales for targeting civilians, which define the intensity of threat to civilians described
above, have remained relatively unchanged throughout history. Protection
of civilians is also likely to remain an objective for military forces as long as
wars are fought amongst the people – regardless of whether operations are
explicitly 'mandated' to do so or not. The areas where future trends will
have most impact are in terms of the 'operational environments' in which
protection will be required and those parameters relating to the ability
and methods with which civilians are attacked and protected. These developments will pose both new challenges and opportunities to find utility of
force to protect in the future.

One trend that may influence the operationalisation of protection of
civilians in the future is growing urbanisation. The fact that 70 per cent of
the world's population will reside in cities by 2050 will inevitably make protection in urban environments more likely, especially in Africa and Asia. A
case in point has been the AMISOM, whose ground operations in Mogadishu were criticised for indiscriminately shelling residential areas from
which al-Shabaab attacked. As a result, AMISOM has successfully implemented an indirect fire policy that establishes no-fire zones in areas of
civilian concentration (Lotze and Kasumba 2012). In other situations, the
difficulty of protecting civilians through air-power alone became apparent
in Libya once fighting moved into the cities, where civilians were used as
human shields and targeted with sniper fire (NATO 2011). In Syria too,
much of the fighting has taken place in densely populated cities, which
presents an even greater challenge. While our own future urban operations are unlikely to resemble the highly destructive urban operations
from our past, as argued in Chapter 8, the perpetrators of violence against
civilians are still likely to follow the traditional model – precisely because
massive destruction serves their purposes of literally crushing all potential
resistance or driving a population out. How protection of civilians can be
operationalised in urban environments, beyond merely restricting the
escalation of force to avoid collateral damage made by own forces, will be
one of the major challenges that must be grappled with in the future.

At the same time, the frequency with which protection of civilians is
required in rural areas may grow in the future. While melting of the ice
has led to speculation about possible conflict between states over access
around the North Pole, the effects of climate change have already led to
conflicts in Africa, particularly over access to water and lands. For
example, in rural areas of Sudan, Uganda and Kenya, pastoralist communities that previously did not interact have now come into conflict as
they are forced to move into new areas where grazing land can be found.
One study has shown that the likelihood of conflict in African countries
increases by 50 per cent in warmer years because of drought – a trend that
would result in 400,000 new battle deaths by 2030 (Burke *et al.* 2009).
Moreover, the proliferation of more modern weapons in these conflicts

has resulted in more sophisticated and destructive military tactics used by both sides leading to more deadly outcomes (Leff 2009). As a result, the need for military protection of civilians may increase in such rural areas. A case in point is the UN peacekeeping mission to South Sudan, which is mandated to protect civilians within an area larger than France and with a smaller urban population (22 per cent) than Afghanistan (CIA 2013).

Some common operational requirements have been identified as particularly critical to the conduct of protection-related tasks in general. These include pre-deployment training for intervening forces on typical protection incidents (e.g. what to do when civilians amass outside bases), flexible force structures that can be detached and operate independently for certain periods, 24-hour operating capability (especially at night), rapid reaction capacity to reach areas where threats become imminent (especially helicopters), and relevant language skills to interact with the population (see Holt *et al.* 2009; Giffen 2011; Kelly and Giffen 2012). Which of these requirements will be most relevant in future operations, however, will depend on the particular threat civilians are under.

In planning for mass atrocity situations, speed is more important than mass, with a premium on capabilities such as transport assets and mobile forces (Sewall *et al.* 2010, p. 18), precisely because perpetrators will have to escalate the violence before intervention occurs or their victims can escape. Boots on the ground are likely to be a critical requirement in most situations in order to deter or confront the military units executing the violence, as exemplified by the inability to prevent ethnic cleansing in Bosnia and Kosovo from the air alone. The exception is perhaps the regime crackdown-scenario, where massive indiscriminate fire intended to coerce a population into submission is only possible through the use of larger and more conventional units, which can more easily be targeted from the air. If a regime decides to expel (ethnic cleansing) or exterminate (genocide) the population instead, irregular and paramilitary units that are much harder to target from the air will be the primary 'executioners' of violence.

The balance in strength between perpetrators and intervening forces is perhaps the parameter where global trends may become most evident. An extreme imbalance of power in favour of the international coalition is considered to be a necessary criterion for a low-cost humanitarian intervention (Pape 2012, p. 58). Here, it is worth noting that the majority of Russia's arms exports go to the most authoritarian regimes in the developing world, such as Sudan, Zimbabwe, Libya (under Gaddafi) and Syria, which also happen to be the type of states that are most likely to use violence against their own people and fail to uphold the R2P principle. As a consequence, military interventions in the developing world are expected to become 'more difficult in the years to come', especially since much of this export includes advanced anti-access platforms and technologies (Bukkvoll 2011, p. 36). Libya was considered to be one of the most likely customers of the most modern version of the S-300 long-range air-defence system before the revolution, and there were plans to modernise the S-125 to the Pechora-2 short-range air-defence system, and possibly even become the first foreign customer of the most advanced of all Russian air-defence systems, the long-range S-400 (Bukkvoll 2011, p. 31).

The global economic recession is also likely to affect the will and capability of the international community to intervene. This prompts the question of whether protection of civilians is possible 'on the cheap'. Arming, training and supporting local rebels or 'self-protection militias' is one alternative – and the former was effective in Libya. Much scepticism surrounds providing such military assistance to both regular and irregular security forces – a military task that has been discussed earlier in Chapter 5. Local forces will always have their own end states, but there may not be many other cheap alternatives given the need for boots on the ground to protect civilians in most situations. Private military companies did play a significant role in reducing violence in Sierra Leone during the 1990s, but this was not without its controversies.

On the positive side, technological advances have generally made it easier to protect civilians and will probably continue to do so in the future. In particular, new weapons associated with the revolution in military affairs have enabled more 'surgical' targeting that has reduced collateral damage significantly. In Kosovo, civilian casualties caused by NATO bombing are said to have been between 489 and 528 (HRW 2000), while in Libya, where virtually all of the 7,700 bombs and missiles used were precision-guided, the figures was fewer than a hundred (HRW 2012). Further advances in technology are likely to continue to provide military forces with new ways of protecting civilians – not simply from own bombs but also from the perpetrators, who are responsible for most casualties anyway.

One example of such protection from perpetrators is the Israeli Iron Dome system, which is designed to counter short-range rockets and artillery shells heading towards populated areas. During fighting in 2012, the Iron Dome is said to have intercepted 90 per cent of the rockets fired from the Gaza strip (TIME 2012). These systems could in theory be used to defend population centres under similar attacks, like Sarajevo during the Bosnian war, or refugee camps, like the ones on the Turkish side of the Syrian border, where NATO has deployed more expensive Patriot missiles designed to intercept ballistic missiles and warplanes. The use of drones for surveillance purposes is something that UN commanders have said would help its protection ability in the DRC and South Sudan (UN 2012). The use of *armed* drones could also provide a low-risk option with which to strafe or strike armed columns of rebels advancing on population centres, or simply for purposes of demonstrating firepower for deterrent effects. On the whole, new low-risk weapons systems may be able to stem the lack of firepower that has largely been missing from UN-operations since Western forces reduced their contributions in the 1990s.

Finally, the spread of information technology may enable civilians to defend themselves better. The ability to document atrocities as they happen was an important factor in precipitating outside help in Libya. By contrast, what happened in Rwanda in 1994 was hardly documented at the time – if the victims there had been able to disseminate pictures as it unfolded on the scale that images flow from current conflicts, the pressure to intervene would have been greater. The long-term effects, however, are less certain given the reluctance to intervene in Syria – yet images from Syria are still a principal source of support for the revolution in the Arab world. At the very least, new information technology, especially mobile phones and the internet, has enabled civilians to provide early warning to local peacekeepers and each other about impending attacks (Kjeksrud and Ravndal 2010). This is an area where cheap solutions can provide effective ways to protect civilians also in the future.

Conclusion

Predicting – at least with any degree of accuracy – what military operations will look like in the future is fraught with uncertainty. Operations where protection of civilians is an objective are no exception. What we *do* know is that the strategic importance of protecting civilians is unlikely to go away unless warfare itself changes first, precisely because its emergence is a result of that change and because perpetrators are unlikely to stop targeting civilians. What is likely to change – and this is where the global trends will be most noticeable – are the means and methods by which perpetrators attack civilians and how protection can be operationalised in response. In particular, the proliferation of more modern weapons and the availability of new technology is likely to influence the balance of strength between perpetrators and both civilians and a potential intervening force. Urbanisation will make urban protection operations an even greater challenge than today, while climate changes may make protection of civilians in rural areas required more often. Technological advances are likely to further reduce the risks of causing collateral damage during operations but may also enable more direct protection of civilians from perpetrators, which is where the biggest potential lies in saving more lives.

The central question, also in the future, will be 'how' military force can best be used to protect civilians in different situations. More than with any other objective, the key lies in 'knowing your enemy', which in this case are perpetrators that deliberately target civilians. As a result, and because protection of civilians may become an objective in operations across the entire conflict spectrum, operation-specific doctrines intended to achieve other types of objective are an unviable source of answers to this overarching question.

Identifying what works, however, may in turn create new dilemmas in the future. First, even when protection of civilians is the primary objective of an entire operation, the ways in which utility of force to protect can be found may be politically unacceptable. In Libya, the expectation was that protection of civilians from Gaddafi's regime would only involve defensive use of force, in line with the mandate that stipulated a no fly-zone, defence of Benghazi and enforcement of the arms embargo. However, it soon became evident that more offensive use of force would be required to reduce the threat more permanently, especially since other instruments of power (economic sanctions, diplomatic efforts and indictments) had failed to do so. When this effectively amounted to 'regime change', the operation understandably drew a lot of criticism. In the future, it may be impossible to get a mandate authorising a similar type of military response to protect civilians.

Yet, ironically, NATO's operation over Libya in 2011 was perhaps the most successful attempt at protecting civilians using military force to date. It effectively removed the threat of physical violence facing the civilians in Benghazi, whose fall could have resulted in a bloodbath; it eliminated the threat from military units firing indiscriminately at population centres; and it eventually removed the principal threat to civilians more permanently by helping the fall of Gaddafi. The paradox is that such an outcome is likely to be required precisely in such cases where the principle of R2P is most relevant (when governments fail to protect civilians or are themselves responsible for attacking them).

Second, in operations where protection of civilians is only one of several objectives, the ways civilians can be protected may clash with how force is required to achieve a different task, for example defeating insurgents. Counter-insurgency operations are often designed to 'clear' insurgent strongholds first. From a protection perspective, this does not make sense: these are the areas where civilians are least likely to be targeted by insurgents because they already have control. Areas where insurgents are losing control, on the other hand, which requires them to threaten and harm civilians, would be more useful to secure first. Similarly, the tactic of provoking fire-fights in order to flush out and eliminate insurgents will not be useful in terms of reducing the threat to civilians, since it is likely to cause more harm than otherwise would have occurred as civilians were not the insurgents' primary target. Thus the tactics required to achieve other objectives will not necessarily be compatible with those through which most utility of military force to protect is found. How to resolve these dilemmas will have to be found on a case-by-case basis, but such insights will not emerge without a basic understanding of how military force can or cannot be used to protect civilians in the first place.

References

Army Times (2010). *McChrystal: Civilian Deaths Endanger Mission*, 30 May. www.army-times.com/news/2010/05/military_afghanistan_civilian_casualties_053010w.

Beadle, Alexander William (2010). "Protection of Civilians in Theory: A Comparison of UN and NATO Approaches". *FFI-rapport 2010/02453*. Kjeller: Norwegian Defence Research Establishment (FFI).

Beadle, Alexander William (2011). "Finding the 'Utility of Force to Protect': Towards a Theory on Protection of Civilians". *FFI-rapport 2011/01889*. Kjeller: Norwegian Defence Research Establishment (FFI).

Beadle, Alexander William (2014). "Protection of Civilians – Military Planning Scenarios and Implications". *FFI-rapport 2014/00519*. Kjeller: Norwegian Defence Research Establishment (FFI).

Beadle, Alexander William and Kjeksrud, Stian (2014). "Military Planning and Assessment Guide for the Protection of Civilians". *FFI-rapport 2014/00965*. Kjeller: Norwegian Defence Research Establishment (FFI).

Bukkvoll, Tor (2011). "Russian Arms Export to the Developing World'. *FFI-rapport 2011/01918*. Kjeller: Norwegian Defence Research Establishment (FFI).

Burke, Marshall B., Miguel, Edward, Satyanath, Shanker, Dykema, John A. and Lobell, David B. (2009). "Warming Increases the Risk of Civil War in Africa". *Proceedings of the National Academy of Sciences of the United States of America* 106(49), 20670–20674.

Christensen, Stian Nordengen (2011). "Approaching Counterfactuals in History, with a Case Study of the Historiography of the War in Bosnia-Herzegovina in 1993". Dissertation for Doctor of Philosophy Degree, University of Oslo.

CIA, *The World Factbook*. www.cia.gov/library/publications/the-world-factbook/fields/2212.html#od, accessed 18 January 2013.

Counterinsurgency Field Manual 3–24, 2006. Boulder, Colo: Paladin Press.

Feil, Scott R. (1998). *Preventing Genocide: How the Early Use of Force Might Have Succeeded in Rwanda*. New York: Carnegie Corporation.

Ford Institute for Human Security (2009). *Protecting Civilians: Key Determinants in the Effectiveness of a Peacekeeping Force*. Pittsburgh: University of Pittsburgh.

Giffen, Alison (2010). *Addressing the Doctrinal Deficit*. Washington, DC: The Henry L. Stimson Center.

Giffen, Alison (2011). "Enhancing the Protection of Civilians in Peace Operations: From Policy to Practice. A background paper prepared for the Australian Government's Asia Pacific Civil–Military Centre of Excellence". 24–26 May 2011. Available at: http://civmilcoe.gov.au/wp-content/uploads/2011/10/Giffen-CMAC.pdf, accessed 18 January 2013.

Global Centre for the Responsibility to Protect (2012). *The Relationship between the Responsibility to Protect and the Protection of Civilians in Armed Conflict*. Policy Brief, http://responsibilitytoprotect.org/The%20Relationship%20Between%20R2P%20and%20POC%20in%20Armed%20Conflict%20Brief%2021%20June%202012.pdf, accessed 18 January 2013.

Gow, James (1997). *Triumph of the Lack of Will: International Diplomacy and the Yugoslav War*. London: Hurst.

Gow, James (2006). "The new Clausewitz? War, Force, Art and Utility – Rupert Smith on 21st Century Strategy, Operations and Tactics in a Comprehensive Context". *Journal of Strategic Studies* 29(6), 1151–1170.

Greenberg, S. B. and Boorstin, R. O. (2001). "People on War: Civilians in the Line of Fire". *Public Perspective*, November/December 2001.

Holt, V., Taylor, G. and Kelly, M. (2009). *Protecting Civilians in the Context of UN Peacekeeping Operations*. New York: DPKO and OCHA.

Human Rights Watch (2000). *Civilian Deaths in the NATO Air Campaign* 12(1). www.hrw.org/reports/2000/nato/Natbm200–01.htm, accessed 18 January 2013.

Human Rights Watch (2001). *Under Orders*. www.hrw.org/sites/default/files/reports/Under_Orders_En_Combined.pdf, accessed 18 January 2013.

Human Rights Watch (2012). *Unacknowledged deaths*. 14 May 2012. www.hrw.org/node/107087/section/2, accessed 18 January 2013.

Iraq Body Count (2010). *Iraq War Logs: What the numbers reveal*, 23 October 2010. www.iraqbodycount.org/analysis/numbers/warlogs/, accessed 18 January 2013.

ISAF Counterinsurgency Guidance, August 2009 Kabul: Headquarters ISAF.

Johnson, David E., Mueller, Karl P., Taft, William H. (2002). *Conventional Coercion Across the Spectrum of Operations: The Utility of US Military Forces in the Emerging Security Environment. MR-1494* (RAND Corporation). www.rand.org/pubs/monograph_reports/MR1494.html

Kelly, Max (2010). *Protecting Civilians: Proposed Principles for Military Operations* Washington, DC: The Henry L. Stimson Center.

Kelly, Max and Giffen, Alison (2011). *Military Planning to Protect Civilians: Proposed Guidance for United Nations Peacekeeping Operations.* Washington, DC: The Henry L. Stimson Center.

Kemp, Richard (2011). "Col Richard Kemp's Speech to 'We Believe in Israel' Conference". London, 15 May 2011. *The Jewish Chronicle Online*, www.thejc.com/blogs/jonathan-hoffman/col-richard-kemps-speech-we-believe-israel-conference-london-15-may.

Kjeksrud, Stian and Ravndal, Jacob Aasland (2010). "Protection of Civilians in Practice: Lessons from the UN Mission in the DR Congo". *FFI-rapport 2010/02378* Kjeller: Norwegian Defence Research Establishment (FFI).

Leff, Jonah (2009). "Pastoralists at War: Violence and Security in the Kenya–Sudan–Uganda Border Region". *International Journal of Conflict and Violence* 3(2), 188–203.

Lidow, N. (2010). "Rebel Governance and Civilian Abuse: Comparing Liberia's Rebels Using Satellite Data", presented at the annual meeting of the American Political Science Association, Washington, DC, 2–5 September 2010. Available at: http://cega.berkeley.edu/assets/miscellaneous_files/wgape/19_Lidow.pdf, accessed 18 January 2013.

Lotze, W. and Kasumba, Y. (2012). "AMISOM and the Protection of Civilians in Somalia". *Conflict Trends*, No. 2, 17–24.

NATO (2011). "Press Briefing on Libya". www.nato.int/cps/en/natolive/opinions_78535.htm, accessed 18 January 2013.

New York Times (2012). "NATO sees Flaws in Air Campaign against Qaddafi", 14 April 2012. www.nytimes.com/2012/04/15/world/africa/nato-sees-flaws-in-air-campaign-against-qaddafi.html?pagewanted=all&_r=0, accessed 18 January 2013.

Pape, Robert (2012). "When Duty Calls: A Pragmatic Standard of Humanitarian Intervention". *International Security* 37(1), 41–80.

PKSOI (2013). *Protection of Civilians Military Reference Guide* Carlisle, Pa: Peacekeeping and Stability Operations Institute (PKSOI).

Sewall, Sarah, Raymond, Dwight and Chin, Sally (2010). *MARO: Mass Atrocity Response Operations: A Military Planning Handbook.* Cambridge, Mass: Harvard Kennedy School, Carr Center for Human Rights Policy.

Silber, L. and Little, Alan (1997). *Yugoslavia: Death of a Nation.* London: Penguin Books.

Smith, R. (2006). *The Utility of Force: The Art of War in the Modern World*. London: Penguin Books.

TIME (2012). "Iron Dome: A Missile Shield That Works", 19 November. http://nation.time.com/2012/11/19/iron-dome-a-missile-shield-that-works/, accessed 18 January 2013.

UN (2012). "Press Conference by Peacekeeping Force Commanders", 20 June. www.un.org/News/briefings/docs/2012/120620_DPKO.doc.htm, accessed 18 January 2013.

UN Joint Human Rights Office (UNJHRO) (2009). "Summary of Fact Finding Missions on Alleged Human Rights Violations Committed by the Lord's Resistance Army (LRA) in the Districts of Haut-Uélé and Bas-Uélé in Orientale Province of the Democratic Republic of Congo". *Special Report by the United Nations Organization Mission in the Democratic Republic of the Congo (MONUSCO) and Office of the High Commissioner for Human Rights (OHCHR)*, December 2009.

UNAMA (2009). *Afghanistan: Annual Report 2008*. http://unama.unmissions.org/Portals/UNAMA/human%20rights/UNAMA_09february-Annual%20Report_PoC%202008_FINAL_11Feb09.pdf

UNAMA (2012). *Afghanistan: Annual Report 2011*. http://unama.unmissions.org/Portals/UNAMA/Documents/UNAMA%20POC%202011%20Report_Final_Feb%202012.pdf, accessed 18 January 2013.

UNAMA (2014). *Afghanistan: Annual Report 2013*. http://unama.unmissions.org/Portals/UNAMA/human%20rights/Feb_8_2014_PoC-report_2013-Full-report-ENG.pdf, accessed 21 March 2014.

Vick, Alan J., Moore, Richard M., Pirnie, Bruce R. and Stillion, John (2001). *Aerospace Operations Against Elusive Targets. MR-1398* (RAND Corporation). Available at www.rand.org/pubs/monograph_reports/MR1398.html.

Weiss, Thomas (2010). "Halting atrocities in Kenya". *Great Decisions 2010*. http://kofiannanfoundation.org/sites/default/files/Kenya%20FPA%20Weiss.pdf, accessed 18 January 2013.

Wheeler, Nicholas J. (2000). *Saving Strangers: Humanitarian Intervention in International Society* Oxford: Oxford University Press.

12 Conclusions

New missions, new tasks

Tore Nyhamar and Per M. Norheim-Martinsen

The purpose of this book has been to improve our understanding of how current trends, and the trajectories they may take, will affect future international military operations. We have shown how the megatrends in demography, economics and technology will result in a very different environment for future international interventions than the one in which current military lessons have been learned. In this concluding chapter, we return to the fundamental question of *how* these expected changes will affect the tasks that military forces will have to carry out as they encounter the international military operations of the twenty-first century. As discussed in Chapter 1, the scope of the book is designed to provide the foresight needed to prepare for the challenges and tasks posed by future international military operations. We have not tried to predict what the *next* mission will be. Instead, foresight handles uncertainty by spelling out a number of scenarios or mission types rather than narrowing down to only those that are considered likely at present. The sample of mission types does not need to be exhaustive to provide foresight. But it does need to identify a sufficient number of interesting new challenges to be helpful. We believe that the authors of the nine previous chapters have succeeded in this task. A number of new challenges and tasks have been identified, while some issues reappear in several of the chapters. The fact that they do gives some indication of their importance and relevance. Moreover, it allows these issues to be approached from different angles, enriching our understanding of them by allowing us to see them in different contexts.

A central observation of this volume is that mission types will resemble each other the most at the tactical level. All firefights share certain characteristics, as they are essentially about fire and movement. A related point is most explicitly made by Robert Egnell and David Ucko in their discussion of counter-insurgency: understanding the character of one type of military operation is helpful in understanding other types, because it deepens the understanding of military force as an instrument. In fact, all military operations are species belonging to the same genus, a point that is reinforced by the way in which the different mission types blend into one another. They can represent a distinct phase in a longer conflict, naturally followed by another phase corresponding to another mission type. In this respect, the high-intensity phase and the ensuing counter-insurgency campaigns in Afghanistan and Iraq offer instructive – if negative – lessons. They can represent a particular capability or domain of conflict, such as cyber or special operation forces. As such, they may be relevant across the whole spectrum of mission types. The essential point is that all military operations, although they may vary greatly, belong in the same category.

Also, a common feature that runs through the analysis of all the nine mission types is that what proves possible tends to fall way short of what one hopes to accomplish. General Rupert Smith's point of departure was that Western forces had failed to find utility of force in wars amongst the people (Smith 2007). In the narrative on an imaginary high-intensity operation in Chapter 3, the campaign is depicted as relatively successful, even as two out of three strategic objectives were not met. While Iraq is considered a counter-insurgency success, at least when compared to Afghanistan, the outcome still leaves a lot to be desired. All military assistance operations would want to produce superbly trained, fully accountable troops under international and recipient nation oversight, but training and assistance missions may still be highly useful even when they fall short of such ideal standards. Both in urban and transnational operations, it is desirable to root out all problems, but missions that accomplish significantly less may still be worthwhile undertaking. All UN operations would ideally like to steer a society onto the path to prosperity, but to accomplish just enough to manage a crisis and avoid escalation may still be enough to justify the operation. Ideally, any military operation would like to prevent the killing of any civilians but may take satisfaction in having saved many and prevented total breakdown. For military power to be useful it need not solve every problem; it need only do more good than harm. Military power cannot do everything, but it can still accomplish something worthwhile.

What is needed is a better understanding of what military power can do and what it cannot do. If we understand what military power can do, we are well on our way to finding utility of force. If this book can make some humble contribution to this larger project, we will at least have found some utility in the analysis.

In the following, we revisit the five parameters to identify how expected changes across the nine mission types will affect the tasks that military forces will have to carry out in the future. Taking the findings of the individual chapters as a point of departure, we then move on to discuss briefly some of the questions that we believe need to be resolved to find utility of force in future international military operations.

New tasks, new requirements – military forces for the twenty-first century

A key objective of this volume has been to develop a comprehensive set of generic parameters for analysing emerging tasks and requirements arising from the mission types. The analysis of the nine mission types against the five parameters set out in Chapter 2 offers a rich and unique source of understanding the future challenges confronting military forces in the twenty-first century. In this final chapter, we will therefore revisit and synthesise what the individual authors have found, with the aim of answering three key questions: Will Western force still be able to intervene? Will it be possible to intervene legitimately? And what tasks, challenges and requirements will they experience if or when they do intervene? We commence by revisiting relative force composition and strength, as this is the parameter that makes it possible for Western forces to intervene. We then move on to discuss consent, the parameter deciding whether intervention is legitimate. The mandate, in turn, decides what Western forces can legally do in the name of international society. The final two parameters, conflict intensity and operational environment, determine what the intervention force encounters once deployed.

Relative force composition and strength

The present Western intervention fatigue is largely caused by the experiences with the large, ambitious and, above all, protracted stabilisation and state building operations in Iraq and Afghanistan. Nearly all of the authors find that the need to avoid operations such as these will greatly influence considerations on future missions. But then again the pendulum may have swung too far in discarding military power as a useful instrument for the West. Chapter 1 concludes that the demand for intervention is still there. Western technological superiority and quality of forces will often still be sufficient for successful intervention in the future, including in high-intensity operations. This contradicts much present gloom about the utility of military force. The chief threat to the imbalance favouring Western forces seems to be the acquisition of anti-access platforms and technology, as described by Alexander Beadle in Chapter 11. However, within the timeframe adopted in this book, it will only be states which will have the necessary capacity to buy and operate these, so that only high-intensity missions and protection of civilians against regime crack-down will be mission contexts in which this will become a major obstacle to Western intervention. Moreover, a trend towards more limited objectives in international military operations does not necessarily mean that the means will be more limited. The chapters on high-intensity operations and on the UN both reject this linear relationship, suggesting that less ambitious objectives may, in fact, be compatible with more use of force in the future.

Faced with a future in which more force may be used in international operations, it is necessary to retain capabilities for the high end of the force spectrum. As stated in Chapter 3, it appears that military operations over and above a certain level of intensity and complexity are better left to a dedicated security organisation or alliance, such as NATO. A permanent military organisation comes with a trained and functional command structure with an agreed working language, standard operating procedures, established doctrine and forces that are interoperable and have trained together. NATO forces operate on the basis of a largely common doctrine, taught in military academies and staff colleges and serving as an intellectual framework for operational planning and execution throughout the Alliance. NATO's capacity to muster joint operations is also a key asset that needs to be retained. Experience shows that, although airpower may yield results in the early phases of a campaign, there will follow phases where a combination of air/ground or air/sea/ground capabilities is needed.

Counter-insurgency operations are a mission type that has proved extremely demanding in terms of numbers of troops. Our analysis does not reveal anything to suggest that this will not continue to be true for future counter-insurgency. The implication for expeditionary counter-insurgency forces is, as pointed out in Chapter 4, that they must not only ensure sufficient capacity but also learn how to identify and exploit opportunities for local partnerships, raise and develop local security forces and find common objectives around which cooperation is possible.

Finally, in the event that Europeans return to UN peacekeeping on a broader scale, the future force composition and strength of UN operations will be very different from what they have been used to. Rest assured, if some of the world's most potent military actors were to put UN boots on the ground again it would undoubtedly be an asset, particularly if future operations need to use more force. But there is also the concern that UN operations and doctrine have evolved, and that current troop contributors have a better understanding of what that means. As stated in Chapter 7, future troop contributors need to make efforts to learn and understand UN peacekeeping through the eyes of the current major troop contributors. These states possess invaluable experience from some of the most challenging conflicts, on the African continent in particular. States such as Brazil have also taken on lead nation responsibilities, proving their worth in confronting challenges of which Western states have had little experience, such as urban peacekeeping, as described in Chapter 8. In addition, these states will remain the dominant providers of ground troops for the foreseeable future. Many European forces, on the other hand, will arrive with recent experience from Afghanistan, only to be met with a quite different modus operandi than they are used to.

Consent

Overall, the issue of consent, which ultimately decides the legitimacy of an operation, has become more fragmented and therefore harder to manage for the intervening force. Military forces in modern international operations require consent not only from state authorities but also from non-state actors. This is perhaps most pronounced when considering the cyber domain. Technological developments have created new arenas for the battle of consent such as You Tube, Facebook, and Twitter. Tomorrow's military forces need to be active in these arenas in order to manage consent in all its forms.

In the more traditional understanding of consent, Chapter 7 points to how, in future UN operations, Blue Helmets will probably continue to operate with the consent of a host nation but with a clearer understanding that, if the host nation forces target civilians intentionally, the UN forces' responsibility to protect will outweigh concerns over keeping the consent of the host nation. In more direct terms, in the future the UN cannot afford to be seen as supporting some of the worst perpetrators of violence against civilians.

When it comes to the management of consent in counter-insurgency operations, contemporary doctrine still presumes a sufficient harmony of interest between intervening and host-nation governments or at least an ability to push the latter toward the 'correct' course of action. Actual practice provides a more sobering perspective, as illustrated in Chapter 4. First, winning legitimacy and support from the local population is inherently difficult for a foreign force, not just because of the readily exploitable symbolism of occupation but also because it is ultimately the legitimacy of the host-nation government – not of the foreign forces – that is at stake. Second, this level of commitment also means that friction between the external counter-insurgents and the host government is inevitable – not only in terms of the vision of the process, but also because of the limited sense of sovereignty that inevitably arises from the long-term deployment of foreign troops.

As in counter-insurgency, the consent of the population is likely to remain the centre of gravity in future urban operations. Chapter 8 describes a situation where negative developments may accumulate over time but where a natural disaster, riots or a massive inflow of refugees may trigger a situation where events spiral out of control and an international intervention comes into question. In such events, the intervening party may have to deal with local actors whose willingness to cooperate is limited, as there may be divergences between local and central authorities, the latter being ultimately the consenting party to a foreign intervention.

Finally, as discussed in Chapter 9, the consent of regional organisations and neighbouring states may be helpful for an intervening force. Regional actors and institutions know the neighbourhood and may share culture, languages and local politics. Local actors will be part of local politics and may exercise influence through personal relationships or control of resources, which may help the intervention. Since they are the most affected by the conflict, they have a legitimate interest in containing it. Furthermore, regional institutions provide legitimacy, as an intervention can hardly be portrayed as Western meddling when it comes as a response to a regional request to deploy military forces. In sum, regional consent will become more important. In addition to being politically desirable and indeed, at times, necessary to obtain a mandate, in a high-intensity setting operational demands make the consent of a neighbouring country to act as a staging area necessary. This shows how regional consent also has a practical dimension.

Mandate

A key trend in international military operations has been a development towards more robust mandates. However, as Gen. Diesen points out in Chapter 3, future military forces may not necessarily be equipped and trained to carry out these mandates. In his analysis of future UN peace-keeping operations, Kjeksrud reveals a similar concern about the development of a potential gap between what the mandate allows and what some of the troop-contributing countries are willing and able to do. Chapter 7 describes a future in which the UNSC moves towards more limited yet more robust peacekeeping mandates. In these operations, UN troops will take sides and, at times, operate without the consent of the main parties to the conflict. At the same time, they will be faced with violent opposition to the UN's use of force. The upside of such a development would be less ambitious – and more realistic – UN mandates. But this may paradoxically lead to more use of force rather than less, as poorly trained and equipped troops are tasked with carrying out what is probably best described as 'higher-intensity' peacekeeping than has been the case in the past.

At the other end of the spectrum we find special operations forces, which often operate without an explicit – or even implicit – mandate but with robust force when needed. Johansen describes how special operation forces have been employed to good effect against the Taliban in 2001. The killing of Osama bin Laden was another notable success, albeit of a very different kind, while special operations forces played a vital role in Operation Unified Protector in Libya. The latter two examples in particular serve to illustrate the murky legal basis for some of the ways in which special operations forces are currently being used. Whereas these operations used to be covert actions, many have now come to regard them as a strategic asset that can be used to great effect – and without the legal constraints, it seems, within which 'conventional' military forces operate. The concept of 'persistent presence' is an attempt, first, to substitute for a large and permanent US military presence and, second, to ameliorate the limited numbers of special operation forces. The idea is to build a 'global special operations forces network' that would deter threats and, if they materialise anyway, to have a capable local force alongside which the Western (i.e. US) forces would fight. Tempting as this option may appear, however, it does represent obvious legal challenges, as these forces will almost always be operating without a clear mandate.

There are situations in which an international mandate is easily obtained, or even redundant. In operations where the host nation requests military assistance, no further mandate is needed. New urban operations may come about in partially failed megacities. In such situations a still functioning state would be the authority to mandate the operation, rendering international approval redundant. But, if a UN mandate is deemed necessary, host-nation consent to the operation would also be helpful. In transnational conflicts, on the other hand, the international legal framework has been the major obstacle to mandates that allow international forces to function properly. However, there are recent examples of a more regional approach to international military operations, especially in Africa.

Indeed, the regional level will generally become more important for future international military operations. The UN has recognised the challenges of regional conflicts from its inception. The UN Charter's Chapter VIII on Regional Arrangements encourages the member states to utilise 'regional arrangements or agencies for enforcement action under its authority'. Moreover, regional organisations may, as described in Chapter 9, provide the necessary legitimacy to obtain an international mandate from the UNSC. China firmly supports the norm of territorial respect but tends to judge issues on a case-by-case basis. Russia, whose view of international intervention tends to be even more restrictive, voted together with China in favour of UNSC Resolution 2098, authorising the Intervention Brigade in DRC, following regional support.

Chapter 11 shows how the principle of R2P, endorsed by UN member states at the World Summit in 2005, remodelled the idea of state sovereignty from strict non-interference to include a state's responsibility to protect its own population from mass atrocities. When states fail to do so, or attack civilians themselves, this responsibility is transferred to the international community, with the use of military force as a last resort. When NATO intervened to protect civilians in Libya in 2011, it was the first time military intervention was authorised with reference to R2P. However, the gradual expansion of targets drew criticism from Russia in particular, which felt that the initial mandate to protect civilians had been hijacked and used to enforce regime change in Libya. To obtain future mandates, the question is whether emerging powers such as China, Russia, India and Brazil will support such counter-regime operations in the future.

However, the point made by Alexander William Beadle in Chapter 11 is worth reiterating. In fact, protection of civilians as a future task in military missions may not be dependent on future R2P mandates. That is, R2P and protection of civilians share a normative foundation – 'the protection of individuals' – but they are concerned with fundamentally different aspects of achieving this goal. R2P is about 'when' and 'under what conditions' it is right to intervene to save civilians. Protection of civilians is concerned with 'how' civilians on the ground can be made safer and better protected, regardless of the reasons for launching an operation. As such, protection of civilians has been a legal obligation for decades, and is what the ICRC and humanitarian organisations do on a daily basis. This is why it will also remain a priority in UN peacekeeping operations, regardless of what happens to R2P.

Finally, there are new mission types that lack precedents in the international legal framework. In Chapter 10, Siw Tynes Johnsen presents some new challenges in the cyber domain that need to be resolved in the future. First, the line between civilian and military targets becomes blurred in the cyber domain. An international consensus seems to be emerging that existing international legal regimes do cover the cyber domain, but the law needs to be interpreted in a way that encompasses the specificities of this domain. Future cyber attacks will thus become test cases for international law. While there may never be such a thing as 'cyber war', it is plausible that a cyber attack could *lead to* war. Another important future test case is how the principle of distinction between civilian and military targets, between the civilian population and combatants, is to be interpreted in international law.

Conflict intensity

Future international missions will put great stress on Western soldiers. They will retain their superiority in training and technology, organisation and operative concepts in order to dominate the battlefield. However, in the future, smaller margins of technological superiority and increased skill levels among adversaries could mean going into combat with smaller margins for error and increased risk of casualties. NATO is the only international institution capable of undertaking a high-intensity campaign, largely thanks to its shared intellectual foundation. But the organisation's operational concepts most relevant to African operations remain underdeveloped. Writing on future high-intensity and UN operations respectively, both Diesen and Kjeksrud note that less ambitious objectives do not necessarily lead to lower conflict intensity. As protection of civilians threatened by their own regime becomes another rationale for high-intensity operations, one may see the contours of a growing gap between the capacity and need for carrying out high-intensity international military operations in the future.

That said, conflict intensity remains first and foremost an intense personal experience for the individual soldier, an experience that in some ways has become more demanding. Egnell and Ucko argue that more force was used in historical examples of counter-insurgency than often assumed, while many past counter-insurgency practices are no longer legal or legitimate. In the future, counter-insurgents may still need to use force, presenting them with intense personal experiences of combat at the tactical level, but they may have to do so within more restrictive rules of engagement. As a result, each soldier will be faced with ever tougher moral decisions.

Another example is found in the contrast between traditional and future urban conflict. As described in Chapter 8, while the destructive effects associated with traditional urban warfare tend to more or less shut cities down, this may not be possible or desirable in tomorrow's mega-cities. Future urban operational concepts will need to find a way to reconcile conflict at the tactical level with the need to retain the city as a functioning organism. Future urban conflicts may be characterised by low overall conflict intensity in military terms but high conflict intensity at a tactical and individual level, as the urban conflict environment poses particularly demanding challenges for the soldiers on the ground. Transnational operations face a similar dilemma, as the intervening force needs to reconcile the use of force at the tactical level with the need to avoid interrupting legitimate economic activity in interconnected areas. Chapter 9 shows how at the low end of such operations soldiers may be faced with criminal gangs, consisting of armed, desperate and violent individuals with a low threshold for violence against anyone threatening their activities. Moreover, criminals, be it in an urban or rural environment, will attempt to hide their activities by blending in with the civilian population, leaving the international force with dilemmas of collateral damage limitation, force protection and effectiveness in fighting small violent groups. In sum, future urban and transnational conflicts may be characterised by overall low conflict intensity but be accompanied by increasingly tough moral choices for the soldiers on the ground.

Interestingly, cyber operations constitute a whole new form of conflict intensity. Conducted on their own, cyber operations would be considered low-intensity warfare, as they have not yet crossed the border into lethality. As such, cyber operations represent a new way of applying political pressure without resorting to physical attacks. However, when applying cyber means in concert with other means of force they can certainly increase conflict intensity. By providing a new front, cyber operations could be one of the building blocks in an effort to produce a new maximum of violence.

Finally, in most operations 'conflict intensity' is defined by the threat posed to our own forces. However, in Chapter 11, Beadle shows how it also needs to be defined by the intensity of threat posed by the perpetrator against civilians. There are enormous variations in terms of how much and in what ways violence is used against civilians. Civilians may be under threat of being massacred, expelled, randomly targeted, selectively killed, brutally punished or looted. Each of these threats will require different military responses.

Operational environment

Terrain matters. Since the Vietnam War, when Western forces have deployed out of area it has often been in desert landscapes, which suited Western forces designed to fight a Soviet invasion. Naturally, they find, for example, jungle terrain, dense forests, lack of infrastructure and tropical climates less familiar. Regardless of physical terrain, however, future international operations will be population-centric. The objective of the operation will be to influence people in the area of operations. This should not be seen as a task separate from military tasks. On the contrary, distinguishing between kinetic operations with the purpose of defeating the enemy and activities with the purpose of gaining local support and legitimacy is unfortunate. All aspects of operations influence the local population. Military organisations are experts in the management of violence to accomplish their objectives. Hence, the use or the threat of force should be seen as the primary tool to influence the local population.

Population-centric operations – that is *all* operations – will take place in all kinds of physical terrain. In addition to the known military challenges associated with operating in deserts, jungles, mountains and so on, comes the fact that the physical terrain also shapes the human terrain. Transnational operations are most likely to take place in inaccessible terrain, be it jungle, mountains or swamps, as in such areas government authority is at its weakest and neighbouring countries are near. This makes reaching the population there harder, but no less important, for the intervening force. The main novelty in new urban operations is that they, in very broad terms, have moved from using conventional high-intensity tactics to counter-insurgency tactics. In other words, they have become more population-centric. The need to influence the population also drives coordination between military and civilian means in UN operations. If or when Western powers do return to UN operations after Afghanistan, they will add valuable enabling capacities that the UN lacks. But they may also, as Kjeksrud points out, find it challenging that the UN has, in fact, come a long way conceptually towards integrating civilian and military means, especially as they will then need to work alongside present troop contributors, which are used to the practical and conceptual tasks inherent in modern UN operations.

Protection of civilians is one of the key tasks. But, as Beadle points out, *how* to protect remains an underdeveloped area of military theory. In order to find utility of force to protect, he argues, we need to improve our understanding of the nature of the perpetrator of violence against civilians. Taking an important step towards refining our understanding of protection of civilians, a typology consisting of seven protection scenarios is defined: genocide, ethnic cleansing, regime crackdown, post-conflict revenge, communal conflict, predatory violence and classic insurgencies. Understanding what the perpetrators of violence want becomes a key feature of the operational terrain in these operations. This, in turn, requires good intelligence about the social terrain and a need to fuse military and civilian sources, in order to produce the information that military forces will need to find utility of force in future operations.

A second general observation is that the operational environment has become more complex, both within and across the range of mission types. Increased complexity creates pressure for specialisation. The sheer number of tasks poses the question whether – and how – Western forces can cope with all the conflicting demands put on them. Let us briefly consider some ways to ameliorate these challenges by revisiting the ideas of operational adaptability, new structural elements and new doctrines.

Operational adaptability can be defined as the ability to shape conditions and respond effectively to changing threats and situations with appropriate and flexible actions. Explicitly making flexible adaption a virtue and a task for Western forces is undoubtedly wise. But how one is to develop the skills that enable soldiers to adapt appropriately remains a challenge. It seems clear that solid, basic soldierly skills are the essential fundament. As Johansen makes clear, it is the very high general soldierly skills that enable Special Forces to adapt to different operational environments. Alternatively, one could teach everyone a little bit about everything to give soldiers a range of options that enable them to adapt. The latter route seems less advisable. First, as this book makes clear, the number of possible operations will simply become too large to make adequate specialised training possible. Second, international operations will become increasingly similar at the tactical level, as peace operations become more robust, and counter-insurgency will require the use of force to influence the population. Hence, in the future, basic soldierly skills will serve as the basis for operational adaptability.

Moving on to structural elements, it seems there is little need to acquire capabilities tailored to each specific mission type. The economic and demographic trends described in Chapter 1 suggest that, if military force is to be a useful instrument in the future, costs need to be kept as low as possible. However, there are nuances. First, in nearly all mission types we find that intelligence that helps towards making sense of the social terrain is key to finding utility of force. This means that intelligence units should be reinforced. However, more can be gained by changing methods, notably fusing civil and military intelligence. Second, the capacity for analysing events and developing doctrine needs to be strengthened. There must be competent staffs available to produce new doctrine quickly when needed, which means that staff resources need to be at a sufficient level to work more or less continuously on developing new doctrine. The analyses of the mission types do not typically focus on specialised units or capabilities. Much more important is the way in which forces are operated.

In the end, doctrines are key to how military forces operate. But doctrines are and ought to be rooted in actual experiences. Ideally, one would wish to leapfrog over experience and devise doctrines tailored to tomorrow's challenges as they come to the fore in the analysis of the different mission types. Practically, applying existing doctrine sensibly, and having the capacity and skill to develop new doctrine fast, is what is possible. Having this capacity is an essential part of the intellectual readiness that we describe in Chapter 1 as needed when preparing for the future.

In their analysis of the British experience with counter-insurgency, Egnell and Ucko caution against too much emphasis on doctrine. UK forces struggled with counter-insurgency in spite of an essentially sound doctrine. True, the doctrine was indeed improved to incorporate recent lessons, but the real problems were that doctrine was not taught and shared and, above all, that UK forces lacked a proper strategy (Ucko and Egnell 2013). We cannot emphasise too much the need for sound strategy in all military operations, strategy that encompasses both political and military knowledge and judgment. Let us, therefore, turn to how our analysis may contribute towards this end. What are the concerns that the generals and politicians need to pay attention to in order to devise sound strategy in future international operations?

Strategy in future international military operations

It seems that Western difficulties in finding utility of force in international operations since 1990 are predominantly due to poor strategy. We do not pretend to be able to solve this issue. Our ambitions with this final section of the book are more modest. First of all, there is no shortage of good analysis of what strategy is, what demands sound strategy must meet and how one could go about getting there (Clausewitz 1832 [1984]; Huntington 1957; Janowitz 1960; Howard 1979; Strachan 2006; Schadlow and Laquement 2009). We make no pretence of improving on these insights. However, we find that the real problem is not theoretical. It is to develop sound strategy in different concrete contexts. Building on the analysis of the mission types covered in this book, we hope to add necessary context to what often becomes a very abstract exercise.

The fundamental problem that has bedevilled Western strategy in international operations is that Western societies have not been organised to fight limited political campaigns. Samuel Huntington (1957) tried to combine civilian control and military effectiveness by separating the civilians and the military. His solution was that the civilian leadership at the top of the chain of command decided the political goals for the military campaign, thus retaining *objective control*. The military then took the political goals handed to them and used their expertise in the management of violence to use military means to reach the goals. Military effectiveness was ensured by shielding the military from unnecessary civilian meddling in developing military skills. Janowitz (1960) developed a competing model. He emphasised *subjective control* to retain civilian control of the military. Since the model is based on the socialisation of the military to achieve civilian control, it is also referred to as the sociological model. Clausewitz pointed out that there is hardly any *purely* military advice. Both models try to unite political goals and knowledge of what military power can and cannot do. In routine operational command and in high-intensity warfare, where the goal is total victory over the enemy forces, the political goals may be sufficiently stable to allow the military to conduct appropriate operations to reach them. Be that as it may, this is emphatically not the case with the international mission types discussed in this book.

Contemporary high-intensity warfare serves limited political goals, demanding an extensive dialogue between politicians and the military. Diesen warns starkly in Chapter 3 against political ends couched in general terms – i.e. stability, peace and the rule of law – and calls for a far more detailed political settlement. He points out that Western militaries have learned the hard way that stability is too general a political end to be useful for developing a sound strategy even for high-intensity operations directed against a regime. Strict definitions of stability and the rule of law would also suffer from the fatal weakness that they are simply too ambitious as political goals for many of the world's trouble spots.

The analysis of counter-insurgency states that civil–military coordination is needed to develop sound strategy. When it does not happen, the consequences tend to be dire. In these military campaigns, since they will inevitably fall short of reaching all desirable objectives, the real challenge is to know what is possible. In the separation model described above, the military contribution to strategy is knowledge of how military means work to reach the desired political ends. In the contemporary and future political context for international military operations, the military also needs to contribute what is possible, rooted in knowledge of what military power can and cannot do. The military is simply closer to the reality on the ground and is therefore in a better position to judge what is possible and what is not. This need to know what is possible is also the reason why *all* authors in this book emphasise the need for better intelligence in future operations. Four key points need to be mentioned. First, international military operations are conducted in an intensely political battlefield. Second, there is a need for other types of intelligence to map the human terrain in the area of operations. Third, intelligence analysis needs to fuse civilian and military sources. Finally, intelligence needs to be shared so that it becomes available to all contributors, civilian and military.

In March 2007, General Petraeus quietly moved away from the goal of turning Iraq into a democracy that would transform the Middle East or of turning the country into a dependable US ally. The previous four years had taught the US military that such goals were unattainable. On the ground in Iraq, the goal was arriving at a more peaceful Iraq that would not explode into a regional war or implode into civil war. Petraeus decided to treat the government of Iraq as a party to the conflict rather than as an ally. Emma Sky, General Odierno's political advisor on Iraq, summarised their informal discussions thus: 'It is a failed state with ungoverned spaces in which the government is part of the problem' (Ricks 2009, pp. 155–156). Petraeus believed that stability could only be achieved by lowering American ambitions and by conditional US support for the Iraqi government, moving somewhat towards an arbiter role (Robinson 2008; Ricks 2009, p. 114). Practically, it meant that militia groups outside the government such as the Sons of Iraq were now acceptable to the Americans. These groups were not helpful in building a stable state let alone a democratic one, but the US military had out of necessity adopted a strategy in which reducing violence had become the primary objective. The informal way the strategy was altered illustrates well the role of the military in modern missions:

> This new American sobriety was the intellectual context for the reduction in the goals of the war. This is a controversial point, because that shrinkage has never been announced or even acknowledged. But it was put into practice every day as a smaller, narrower set of aims.
>
> (Ricks 2009, p. 164)

It is what Morris Janowitz calls unanticipated militarism, brought about by 'lack of effective traditions for controlling the military establishment, as well as from a failure of civilian leaders to act relevantly and consistently' (Janowitz 1960, pp. 12, 143). The point is that decisions to find realistic ambitions are forced upon the military because someone has to acknowledge responsibility. If done sooner rather than the later, which was unfortunately the case in Iraq, it will also lead to a more realistic strategy before the situation deteriorates.

The political–military dialogue faces the additional challenge that the limited political goals of the campaign do not necessarily lead to limited military means. More specifically, Diesen warns that a joint and combined campaign is required, as removing one weapons system or service renders the remaining weapons and services militarily less effective. This is the same issue that has become an urgent problem in recent counter-insurgency campaigns. Unfortunately, the fact that the problem is widely recognised does not mean that it has been properly addressed. In high-intensity warfare, it is even easier to slip into the trap that in conventional war only conventional civil–military dialogue is required. This is not the case. For example, operations in sub-Saharan Africa will need another African country willing to serve as staging area, which should not be land-locked and should possess infrastructure that permits the landing of heavy strategic aircraft. Any successful military operation will need to clear this political hurdle.

We also find that the technical nature of military assistance operations may easily lead to the erroneous thinking that the mission can do without a political strategy. However, a first issue that needs to be appreciated for domestic political reasons is that mentoring may involve sending troops in harm's way without the protection that Western forces usually bring to the battlefield. Indeed, for political reasons they may well be targeted during the mission. A second issue is to develop professionalism in the forces that we assist. This is a question that goes beyond installing the right values during training. The security forces, be they military or police, cannot function properly without support from appropriate domestic institutions. This is necessary in order to ensure that they use what they learn in ways that are accountable to the host nation authorities and in the interest of the population. In short, military assistance needs to be part of a larger strategy in order to succeed. It needs to be part of a strategy of state build-ing, no matter how modest, to be successful. This means that if there is any doubt about *who* is the government a decision needs to be made at the political level on whom to support.

One of the temptations of special operation forces is that they seem to offer a purely military solution to irregular threats, a prominent feature of the new security environment. If the problem is to kill a known terrorist, say Osama bin Laden, such forces may indeed be just that. However, taking out targets, even high-value ones, is a short-term policy and not a long-term solution. If the problem is to build government capacity in ungoverned space, special operations forces may be used in a limited tacti-cal role but need a broader political strategy.

In modern urban operations, a sort of tactical counter-insurgency is required in the sense of bringing some kind of governance back into those parts of the city where it has collapsed. The challenge is that, if law and order and basic government services have collapsed, any intervening force needs to have a plan of how to restore these basic government functions in order to be successful. In transnational operations, a successful strategy needs to prevent undesirable flows of communication while allowing desir-able ones. This requires a strategy for both sides of the border. But there are ways of doing this other than operating international forces on both sides of the border.

As a new domain, cyberspace can appear less regulated and more diffi-cult to understand and predict than other domains. However, it does offer a way to conduct less harmful operations than using conventional military force. A problem is that it can offer 'plausible deniability' to the attacker. For example, Russian authorities denied responsibility for the cyber opera-tions against Estonia and Georgia in 2008, blaming patriotic hackers, even though the attacks were traced back to computers on Russian soil. In the latter case, cyber operations were concurrent with ordinary military operations, representing what is still the most effective way to carry out cyber attacks. Another challenge posed by the relative immaturity of rules in the cyber domain is that it is neither clear what constitutes a cyber attack nor what it means in terms of alliance obligations.

When it comes to protection of civilians, the most important point to be made is that it has become an objective important enough to warrant strategic and operational thinking in its own right. However, two political issues need to be settled in future operations. In operations where protection is the primary objective, what works to protect civilians may be politically unacceptable for other reasons. The most effective way to protect civilians in Libya led to regime change. In operations where protection of civilians is only one of several objectives, these objectives may clash with each other. In counter-insurgency one would like to clear enemy strongholds first, but this is the least likely place for civilians to be at risk. In situations like these, sound strategy needs to realise the dilemmas involved in making the best choice among the alternatives available.

Finally, even though UN operations may have lower ambitions in terms of state building, there is reason to believe that they will not become less complex. There are certain things that these missions need to accomplish, and they will not go away. A key concern is that the political implications of today's robust mandates may not be fully understood. Becoming more robust and taking sides in intra-state conflicts will also inherently make the use of force more political in UN operations, despite being linked to the ideal goal of protection of civilians. So far, this development has gone by without much notice, which may again leave military commanders at the tactical level with the decisions of when to use force and when to remain passive. On the one hand, the result may be use of force without political support. On the other, there is the disturbing consequence that UN forces may yet again become passive bystanders to atrocities they could have prevented.

References

Clausewitz, Carl Von (1832) [1984]. *On War.* Edited and translated by Michael Howard and Peter Paret. Princeton, NJ: Princeton University Press.

Howard, Michael (1979). "The Forgotten Dimensions of Strategy". *Foreign Affairs* 57(5), 975–986.

Huntington, Samuel P. (1957). *The Soldier and the State: The Theory and Politics of Civil–Military Relations.* Cambridge: Belknap.

Janowitz, Morris (1960). *The Professional Soldier: A Social and Political Portrait.* New York: Free Press of Glencoe.

Ricks, Thomas E. (2009). *The Gamble: General David Petraeus and the American Military Adventure in Iraq.* New York: Penguin Press.

Robinson, Linda 2008. *Tell Me How this Ends: General David Petraeus and the Search for a Way out of Iraq.* New York: PublicAffairs.

Schadlow, Nadia and Richard A. Laquement (2009). "Winning Wars, Not Just Battles: Expanding the Military Profession to Incorporate Stability Operations". In Suzanne C. Nielsen and Don M. Snider, Eds. *American Civil–Military Relations: The Soldier and the State in a New Era*. Baltimore, Md: Johns Hopkins University Press.

Smith, Rupert (2007). *The Utility of Force: The Art of War in the Modern World*. New York: Knopf.

Strachan, Hew (2006). "Making Strategy: Civil–Military Relations after Iraq". *Survival* 48, 59–82.

Ucko, David H. and Robert Egnell (2013). *Counterinsurgency in Crisis: Britain and the Challenges of Modern Warfare*. New York: Columbia University Press.

Index

Page numbers in *italics* denote tables.